THE PLANT SEED: DEVELOPMENT, PRESERVATION, AND GERMINATION

ACADEMIC PRESS
RAPID MANUSCRIPT REPRODUCTION

The Proceedings of a Symposium on the
Development, Preservation, and Germination
of the Plant Seed Held at the University of Minnesota,
St. Paul, Minnesota,
March 21–23, 1978

The Plant Seed: Development, Preservation, and Germination

edited by

IRWIN RUBENSTEIN
Department of Genetics and Cell Biology

RONALD L. PHILLIPS
CHARLES E. GREEN
B. G. GENGENBACH
Department of Agronomy and Plant Genetics

University of Minnesota
St. Paul, Minnesota

ACADEMIC PRESS:

A Subsidiary of Harcourt Brace Jovanovich, Publishers
New York London Toronto Sydney San Francisco
1979

ACADEMIC PRESS, INC.
111 Fifth Avenue, New York, New York 10003

United Kingdom Edition published by
ACADEMIC PRESS, INC. (LONDON) LTD.
24/28 Oval Road, London NW1 7DX

Library of Congress Cataloging in Publication Data

Main entry under title:

The Plant seed, development, preservation, and germination.

Based on proceedings of a symposium held at the University of Minnesota, Mar. 21–23, 1978.
Includes index.
1. Seeds—Congresses. 2. Seeds—Preservation—Congresses. 3. Germination—Congresses.
I. Rubenstein, Irwin. II. Minnesota. University.
QK661.P54 582′.0467 79-22179
ISBN 0-12-602050-7

PRINTED IN THE UNITED STATES OF AMERICA

79 80 81 82 9 8 7 6 5 4 3 2 1

Contents

SECTION III
Seed Germination

Contributors

Numbers in parentheses indicate the pages on which authors' contributions begin.

Louis N. Bass (145), USDA, SEA–AR, National Seed Storage Laboratory, Fort Collins, Colorado 80521

R. N. Beachy (67), Department of Biology, Washington University, St. Louis, Missouri 63130

J. Derek Bewley (219), Department of Biology, University of Calgary, Calgary T2N 1N4, Alberta, Canada

F. A. Bliss (3), Department of Horticulture, University of Wisconsin, Madison, Wisconsin 53706

B. U. Buchbinder (3), Department of Horticulture, University of Wisconsin, Madison, Wisconsin 53706

B. Burr (27), Biology Department, Brookhaven National Laboratory, Upton, New York 11973

F. A. Burr (27), Biology Department, Brookhaven National Laboratory, Upton, New York 11973

R. M. K. Dale (129), Department of Biology, Yale University, New Haven, Connecticut 06520

L. S. Dure III (113), Department of Biochemistry, University of Georgia, Athens, Georgia 30602

Michael Freeling (85), Department of Genetics, University of California, Berkeley, California 94720

M. M. Goodman (171), Department of Statistics, North Carolina State University, Raleigh, North Carolina 27607

T. C. Hall (3), Department of Horticulture, University of Wisconsin, Madison, Wisconsin 53706

T. J. V. Higgins (241), Division of Plant Industry, C.S.I.R.O., Canberra, ACT 2601, Australia

Bor-Fuei Huang (203), The Institute for Cancer Research, The Fox Chase Cancer Center, Philadelphia, Pennsylvania 19111

William J. Hurkman (49), Department of Botany and Plant Pathology, Purdue University, West Lafayette, Indiana 47907

J. V. Jacobsen (241), Division of Plant Industry, C.S.I.R.O., Canberra, ACT 2601, Australia

Brian A. Larkins (49), Department of Botany and Plant Pathology, Purdue University, West Lafayette, Indiana 47907

Y. Ma (3), Department of Horticulture, University of Wisconsin, Madison, Wisconsin, 53706

J. T. Madison (67), U.S. Plant, Soil, and Nutrition Laboratory, Ithaca, New York 14853

Abraham Marcus (203), The Institute for Cancer Research, The Fox Chase Cancer Center, Philadelphia, Pennsylvania 19111

R. C. McLeester (3), Department of Horticulture, University of Wisconsin, Madison, Wisconsin 53706

Nina L. Pearlmutter (49), Department of Botany and Plant Pathology, Purdue University, West Lafayette, Indiana 47907

J. W. Pyne (3), Department of Horticulture, University of Wisconsin, Madison, Wisconsin 53706

Shirley Rodaway (203), The Institute for Cancer Research, The Fox Chase Cancer Center, Philadelphia, Pennsylvania 19111

S. M. Sun (3), Department of Horticulture, University of Wisconsin, Madison, Wisconsin 53706

I. M. Sussex (129), Department of Biology, Yale University, New Haven, Connecticut 06520

J. F. Thompson (67), U.S. Plant, Soil, and Nutrition Laboratory, Ithaca, New York 14853

D. H. Timothy (171), Department of Crop Science, North Carolina State University, Raleigh, North Carolina 27607

James C. Woodman (85), Department of Genetics, University of California, Berkeley, California 94720

J. A. Zwar (241), Division of Plant Industry, C.S.I.R.O., Canberra, ACT 2601, Australia

Preface

Over 70% of the world food supply is derived directly from the seeds of a small number of field crops. In most crop species, seed formation initiates the next reproductive cycle. These dual requirements, food and reproduction, place unusual demands on the biological processes of seed development and germination. While plant improvement programs often focus on increased seed yield as a central goal, other factors such as seed quality must also be considered. An understanding of the vast array of physiological, molecular, and developmental events associated with seed development and germination is needed to increase supplies of these grains and to preserve the genetic resources of these species. Preservation and appropriate maintenance of the diverse but dwindling germplasm resources is imperative in order to meet future food needs.

With this in mind a symposium was held in 1978 at the University of Minnesota to discuss the development, preservation, and germination of the plant seed. This represents the fifth symposium at Minnesota that focused on recent and future trends in eukaryotic biology of particular importance to plant improvement. The meeting was attended by approximately 180 people interested in various aspects of the plant seed. Unfortunately, the enthusiasm generated by the presentations and the informal exchanges among those in attendance cannot be captured in print. This book presents for the reader much of the information discussed at the symposium. The editors hope that the bringing together of this diverse information will be useful for the reading audience.

The topics of this volume focus on various aspects of the plant seed. The first group of papers describes genetic, hormonal, and molecular events associated with seed development, with particular attention given to the molecular biology of storage protein formation; the second group of papers examines the physiological and genetic aspects of germplasm preservation. The final group of papers examines molecular aspects of seed germination.

The editors gratefully acknowledge financial support from the National Institutes of Health and the University of Minnesota's College of Biological Sciences, Departments of Botany, Department of Genetics and Cell Biology, the College of Agriculture's Department of Agronomy and Plant Genetics, and the Department of Genetics and Cell Biology; Sigma Xi (national and local); DeKølb AgResearch, Incorporated; Funk Seeds International; Green Giant Company; Northrup King Company; Pfizer Genetics, Incorporated; and Pioneer Hi-Bred International, Incorporated.

Seed Development

THE MAJOR STORAGE PROTEIN OF FRENCH BEAN SEEDS: CHARACTERIZATION *IN VIVO* AND TRANSLATION *IN VITRO*[1]

T. C. Hall
S. M. Sun
Y. Ma
R. C. McLeester
J. W. Pyne
F. A. Bliss
B. U. Buchbinder

Department of Horticulture
University of Wisconsin
Madison, Wisconsin

The mature seed of the French bean (*Phaseolus vulgaris*, L.) contains some 25% protein and is a valuable food source for humans. The economic significance of this crop can be judged from the annual production figures: In the United States and Canada 1 million tons are grown; in Latin America about 4 million tons are grown. The total world production exceeds 12 million tons yearly (Sanders and Alvarez, 1978). We are studying the molecular biology of events occurring during the synthesis and accumulation of seed proteins as a means to identify and understand the function of structural and regulatory genes involved in their expression. The information obtained will be used in the improvement of protein quantity and quality of dry beans through classical genetic manipulation. We hope that it will also permit a realistic evaluation of the potential for novel approaches to crop enhancement based on recombinant DNA and other somatic engineering techniques. This article reviews our progress toward characterizing the major bean seed protein and its biosynthesis.

[1]*Research supported by NSF grant PCM 73-21675, NIH training grant 5 T32 GM7215, the Graduate School and the College of Agriculture and Life Sciences of the University of Wisconsin, and the Herman Frasch Foundation.*

ISBN 0-12-602050-7

BEAN SEED PROTEIN COMPOSITION AND CHARACTERIZATION

The protein of mature bean seeds consists of about 60% globulin (protein requiring an appreciable ionic strength solution for solubilization), 20% albumin (enzymes and other water-soluble proteins), 10% glutelin (protein soluble in alkaline solutions) and 3% prolamine (ethanol-soluble protein). Free amino acids account for some 7% of the total seed nitrogen. The proportions of these constituents vary from cultivar to cultivar (Ma and Bliss, 1978). The globulin component can be subdivided into globulin-1 (or G1), a major fraction that requires high levels of salt for solubility, and the globulin-2 (G2) fraction that is soluble in solutions of relatively low salt concentration. The ratio of G1 to G2 is about 6:1. The bean cultivar Sanilac is widely grown and represents a useful seed line against which other bean cultivars can be compared; it has a good yield and a fairly high protein content. Although more cultivars need to be tested, a comparison of several bean varieties having similar total yields revealed that most of those higher in seed protein than Sanilac contained over 41% G1 protein and that lines having less than 41% of the total seed protein in the G1 fraction were poorer in protein than was Sanilac (Table I).

From Table II it can be seen that the methionine level in G1 protein is low; this is a major factor in the deficiency of this nutritionally essential amino acid in bean protein. For the cultivar Tendergreen, the methionine value corresponds to about 3 methionine residues per polypeptide subunit. However, the methionine content of G1 protein from other seed lines is higher, and it will be useful to determine the molecular basis for this variation.

The development of an acidic extraction procedure for seed globulin (McLeester *et al.*, 1973; Sun and Hall, 1975) was valuable since it permitted complete separation of fraction G1 from G2 (Fig. 1). The G1 fraction from the cultivar Tendergreen is resolved into three subunits (here referred to as α, mol. wt. 53,000; β, mol. wt. 47,000; and γ, mol. wt. 43,000) on dissociation and electrophoretic separation on SDS-polyacrylamide gels. Segregation of two forms of the α subunit, one having a mol. wt. of 53,000 (α53) and the other of 50,500 (α50.5) was shown to occur as expected for control by a single Mendelian gene (Romero *et al.*, 1975; Hall *et al.*, 1977b). Variant forms of the α subunit have also been observed for G1 from other bean cultivars (Fig. 2).

TABLE I. *Analysis of Bean Lines Having High or Low Seed Protein Content*

Seed line No.	Total protein (%)	Content of individual fractions in seed[a]				
		G1	G2	Albumin	Other	Amino acids
		high protein				
742032	32.2	51.2	6.0	11.6	23.6	7.6
742077	29.5	41.2	5.9	14.2	30.1	8.6
742015	27.5	45.9	7.4	11.1	26.3	9.3
742071	26.2	45.3	6.2	13.1	28.9	6.7
742065	24.7	42.6	7.4	15.6	28.1	6.3
$\bar{x}$	28.0	45.2	6.6	13.1	27.4	7.7
		low protein				
742016	24.0	40.8	4.8	14.3	34.8	5.3
742069	21.4	39.9	7.6	15.3	31.9	5.3
742090	19.9	40.1	4.3	14.7	35.0	5.9
742047	19.1	34.4	5.1	19.8	34.7	6.0
742066	18.9	34.4	4.7	17.7	36.9	6.3
$\bar{x}$	20.7	37.9	5.3	16.4	34.7	5.8
		standard lines				
Sanilac	25.6	36.4	7.5	12.4	39.2	4.5
BBL 240	27.7	44.3	7.4	15.5	26.8	6.0

[a]All data are converted from Kjeldahl nitrogen determinations using a factor of 6.25 and are expressed as a percentage of total protein. The fraction designated Other represents protein not soluble in ascorbate-HCl (pH 4.5); it includes prolamines, glutelins, and acid-insoluble albumins. The mean seed yield (gm/plant) was 27.4 for the high and 27.6 for the low protein lines.

TABLE II. Amino Acids in the Gl Fraction of Seed Protein from Several Lines of Phaseolus vulgaris.[a]

Amino acid	Seed line					
	Tendergreen	BBL 240	PI 229815	Sanilac	PI 207227	PI 302542
Ala	3.47	3.01	3.25	3.40	3.38	3.91
Arg	5.42	7.45	7.83	7.43	8.08	8.00
Asp[b]	14.47	12.76	12.64	13.47	13.26	14.84
Cys	---	0.13	0.08	0.13	0.06	0.07
Glu[b]	19.88	17.61	17.11	17.38	17.72	18.04
Gly	3.30	2.63	2.73	2.49	2.41	3.08
His	3.60	3.59	3.13	3.39	3.27	3.88
Iso	5.47	4.68	4.73	4.66	5.10	5.46
Leu	9.42	10.00	10.26	10.84	10.42	11.82
Lys	7.60	7.38	7.77	7.91	7.49	8.87
Met	0.57	0.87	0.95	0.97	0.97	1.36
Phe	7.78	10.82	9.92	9.52	8.78	8.77
Pro	2.61	2.96	3.16	2.08	3.66	3.72
Ser	4.52	4.99	5.11	5.20	4.61	5.23
Thr	2.56	2.81	3.04	2.78	2.91	3.41
Tyr	3.38	3.70	3.82	3.79	3.78	4.53
Val	5.89	4.63	4.47	4.62	4.10	5.01

[a]*Data are corrected to 100% recovery and expressed as mg/100 mg of protein.*

[b]*Values including their amide form.*

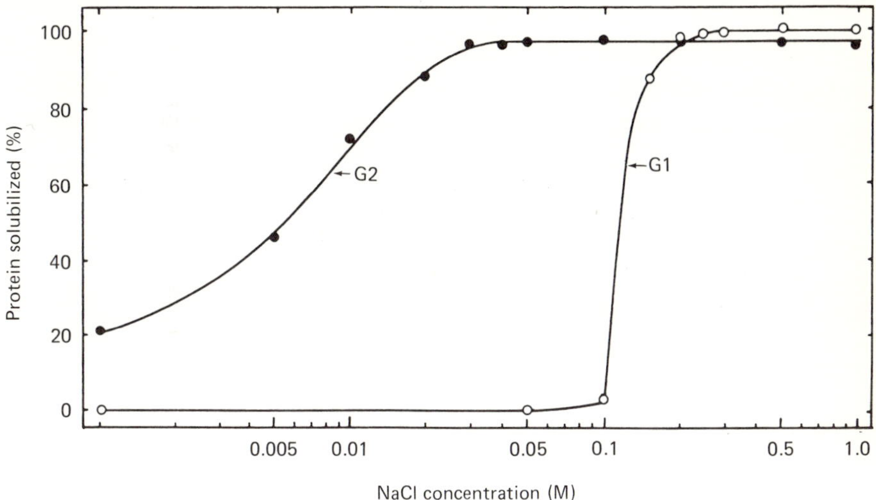

FIGURE 1. Solubility of G1 and G2 proteins as a function of NaCl concentration (from Sun and Hall, 1975; courtesy of the American Chemical Society).

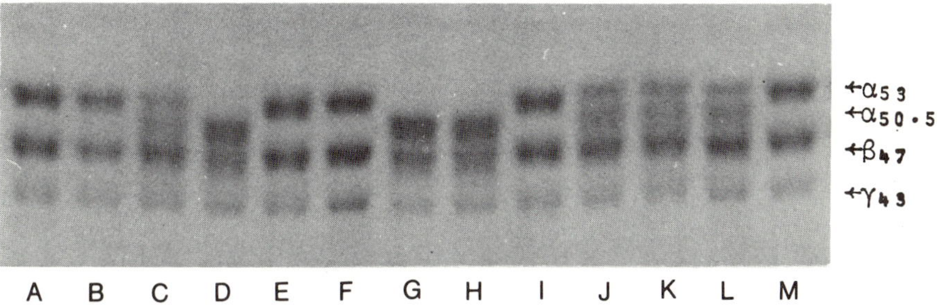

FIGURE 2. Electrophoretic separation of G1 protein derived from several seed lines. Protein from seeds of the cv. Tendergreen was run in lanes A, E, I, and M; from BBL 240 in lane B; from F_1 seed of the cross PI 229,815 x BBL 240 in lanes C and K; from PI 229,815 in lanes D and H; from cv. Canadian Wonder in lane F; and from cv. Seafarer in lane G. Mixtures of equal volumes of extracts from BBL 240 and PI 229,815 were run in lane J, and of Canadian Wonder and Seafarer in lane L. The top and bottom (+ electrode) of the gel are not shown (from Hall et al.*, 1977b; courtesy of the American Society of Plant Physiologists).*

The α, β, and γ subunits of G1 reversibly associate, depending on pH conditions (Sun *et al.*, 1974), to give different conformational forms:

$$\underset{\substack{\text{Tetramer}\\ 18.2\ \text{S}}}{(\alpha\beta\gamma)4} \underset{\text{pH } 5.4}{\overset{\text{pH } 7.4}{\rightleftharpoons}} \underset{\substack{\text{Protomer}\\ 7.1\ \text{S}}}{4(\alpha\beta\gamma)} \underset{\text{pH } 10.5}{\overset{\text{pH } 12.0}{\rightleftharpoons}} \underset{\substack{\text{Peptides}\\ 3.0\ \text{S}}}{4\alpha + 4\beta + 4\gamma}$$

This reaction has been used as the basis for an affinity technique for the separation of G1 protein from the other bean seed proteins (Stockman *et al.*, 1976).

At neutral pH values, the G1 fraction cannot be distinguished from the G2 fraction by analytical ultracentrifugation (Fig. 3). For this reason, protein sedimenting at 7S (but actually a mixture of G1 and G2 fractions) was thought to be the most abundant storage protein of *P. vulgaris* (Danielsson, 1949). This has lead to confusion since some authors consider 7S proteins to be vicilins (Derbyshire *et al.*, 1976), although the original definition (Osborne and Campbell, 1898) designated the pea globulin fraction soluble in relatively weak saline solution as vicilin and that needing strong saline solution for solubilization as legumin. As shown in Fig. 1, G1 requires a considerably stronger saline solution for solubilization than does G2, and by this definition is legumin-like, although its sedimentation properties at neutral pH are vicilin-like. To avoid confusion due to terminology, we have chosen to use an operational definition (G1: the globulin that precipitates first on dilution of an acidic salt extract with water) for the major bean seed storage protein. The subunits of G1 are glycosylated (Hall *et al.*, 1977b), and it appears that G1 is essentially the same protein as that described as glycoprotein II by Pusztai and Watt (1970). However, the procedure described for the isolation of glycoprotein II is much more complex and yields material that is no more purified than that obtained by our simple technique for G1 extraction in acidic saline solution (McLeester *et al.*, 1973; Sun and Hall, 1975). Thus retention of the G1 terminology appears appropriate until some meaningful system for defining legume seed proteins has been agreed upon.

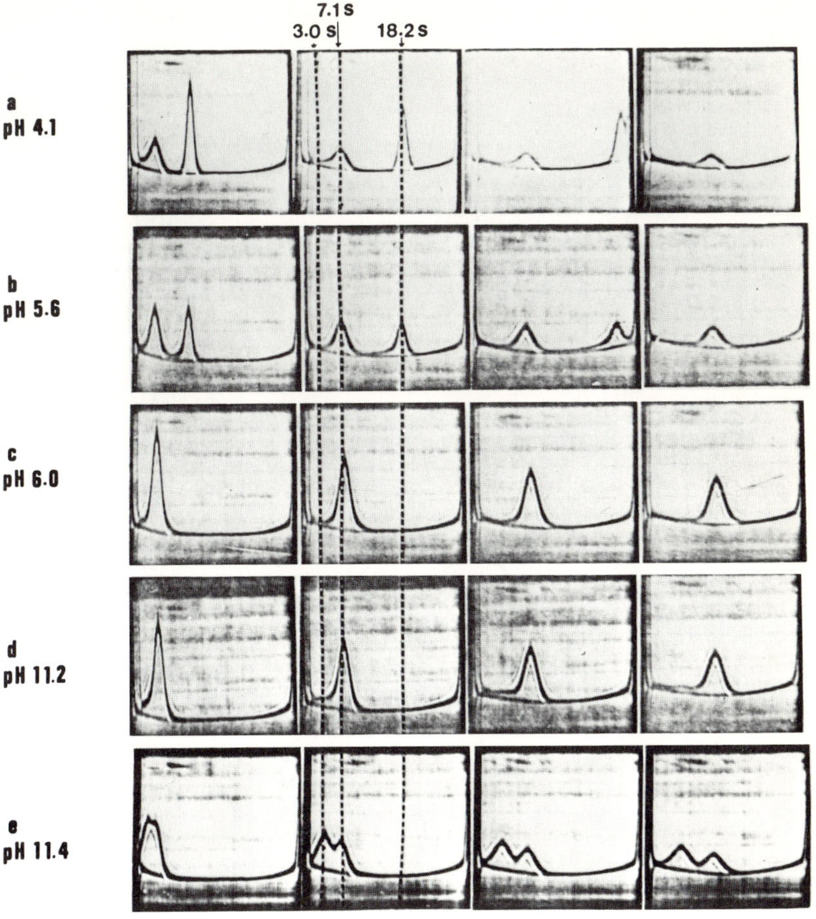

FIGURE 3. Ultracentrifugal patterns of a mixed sample of G1 and G2 globulins. The sample contained 2 mg/ml G1 protein and 2 mg/ml G2 protein. Under acidic conditions (a) two boundaries are discerned (G1 at 18.2S and G2 near 7S); but increasing the pH of the solution dissociated the G1 globulin to the protomeric form, decreasing the amount of material in the 18.2S location (b), until only a single boundary was apparent at about 7.1S (c,d). At pH 11.4, the G1 material dissociated to 3.0S peptides while the G2 globulin remained at approximately the 7.1S location (e).

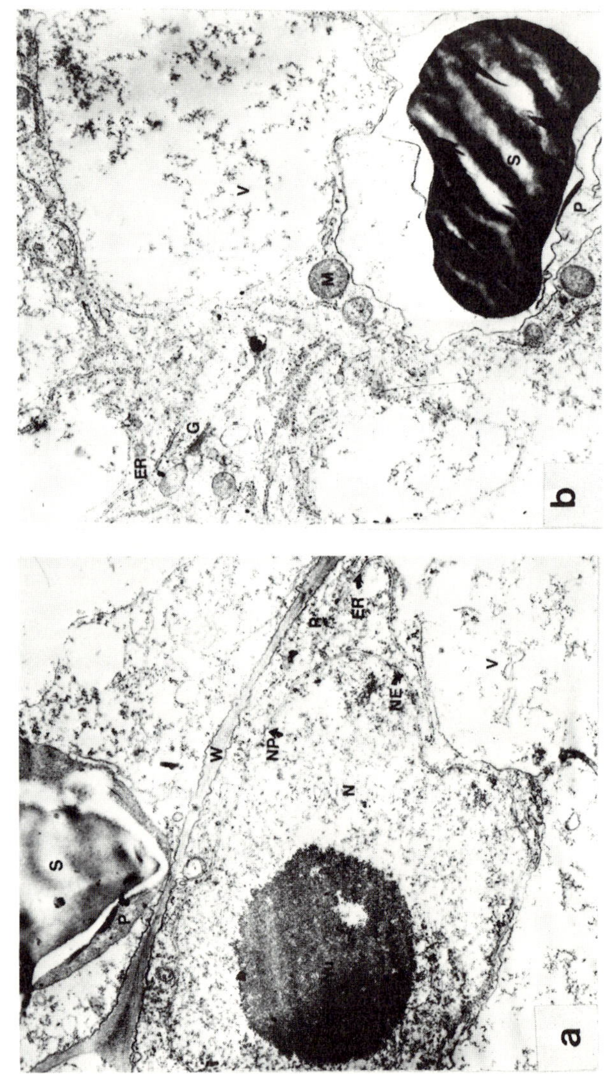

Fig 4 (a) *(b)*

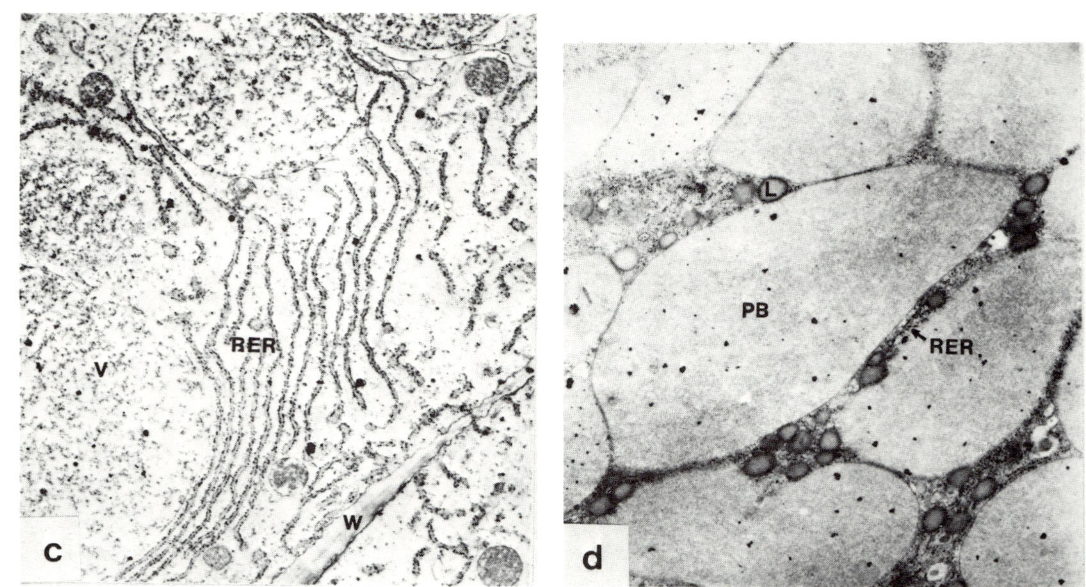

FIGURE 4. Electron micrographs of bean cotyledons during ripening. (a) 5mm seed (x 11,800); (b) 10 mm seed (x 13,000); (c) 14 mm seed (x 9000); (d) 19 mm seed (x 13,000). ER, endoplasmic reticulum; G, golgi body; L, lipid droplet; M, mitochondrion; N, nucleus; NE, nuclear envelope; NP, nuclear pore; Nu, nucleolus; P, plastid; PB, protein body; R, ribosome; RER, rough endoplasmic reticulum; S, starch grain; V, vacuole; W, cell wall.

CYTOLOGICAL CHANGES DURING SEED DEVELOPMENT

Cytological and physiological changes in *P. vulgaris* seeds during development have been followed in several studies (Carr and Skene, 1961; Öpik, 1968; Walbot *et al.*, 1972; Sun *et al.*, 1978). Parameters used as the basis for observing these changes have included seed or cotyledon length and the number of days after flowering (DAF). For many purposes, seed length is an adequate measurement of development.

Electron microscopy of the developing cotyledon revealed that at the 5 mm stage the cells were highly vacuolated, the vacuoles containing little electron dense material. There was relatively little endoplasmic reticulum, but the cells were rich in ribosomes. Many plastids with starch grains and mitochondria could be seen within the cytoplasm. Golgi bodies were present, and the nuclei contained large nucleoli of granular appearance (Fig. 4a). An increase in the amount of rough endoplasmic reticulum was evident by the 10 mm stage (Fig. 4b), and by the 14 mm stage (Fig. 4c) copious amounts of rough endoplasmic reticulum were seen to surround vacuoles. At this stage, the vacuoles were evidently filling with electron-dense material, presumably protein, and appeared to be subdividing. When the seeds had attained 19 mm (Fig. 4d), the vacuoles were packed with protein; they are correctly designated protein bodies at this stage. As in the case of zein-containing protein bodies in corn seeds (Burr and Burr, 1976), the bean seed protein bodies are surrounded by a single unit membrane.

PROTEIN CHANGES IN SEEDS DURING DEVELOPMENT

Electrophoretic analysis of protein changes during maturation of bean (cultivar Tendergreen) seed on SDS gels revealed the presence of G1 polypeptides at the 9 mm seed stage; this was followed by a rapid increase in the proportion of these polypeptides until they dominated the electrophoretic profile (Fig. 5). The relative amounts of the α, β, and γ subunits of G1 protein did not change during development, but the proportions of peptides associated with the G2 fraction did vary. This provides additional evidence for our belief (McLeester *et al.*, 1973) that G1 is a homogeneous protein but that the G2 fraction contains several different protein species.

The complete separation of G1 and G2 proteins made possible by the acidic extraction procedure is especially valuable since the purified protein fractions can be used as antigens for the production of rabbit serum containing specific antibodies.

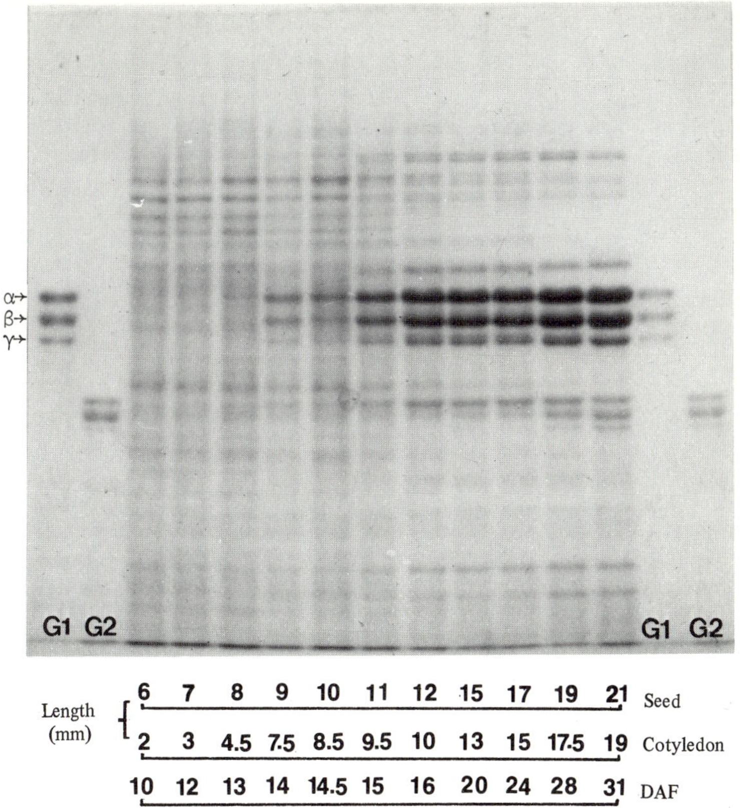

FIGURE 5. Electrophoretic separation (16% acrylamide gel) of proteins extracted from bean cotyledons of different developmental stages. Standard G1 and G2 proteins were run in the outer lanes. The number of days after flowering (DAF), seed and cotyledon lengths are indicated for each extract (from Sun et al.*, 1978; courtesy of the American Society of Plant Physiologists).*

The antibody to G1 has been found to be monovalent, reacting strongly to G1 protein, but showing no immunoprecipitation with G2 or albumin fractions. The G2 antibody shows no reaction against G1 protein. G1-specific antibody has been used for immunoelectrophoretic analysis of G1 content in ripening cotyledons by the "rocket" procedures of Laurell (1967) and Weeke (1973). Figure 6 shows that although very low levels of G1 protein are present in seeds as small as 7 mm in length, active expression of genetic information for G1 protein commences at the 11 mm stage. This is dramatically illustrated by the electrophoretic separation of protein from seeds of increasing age

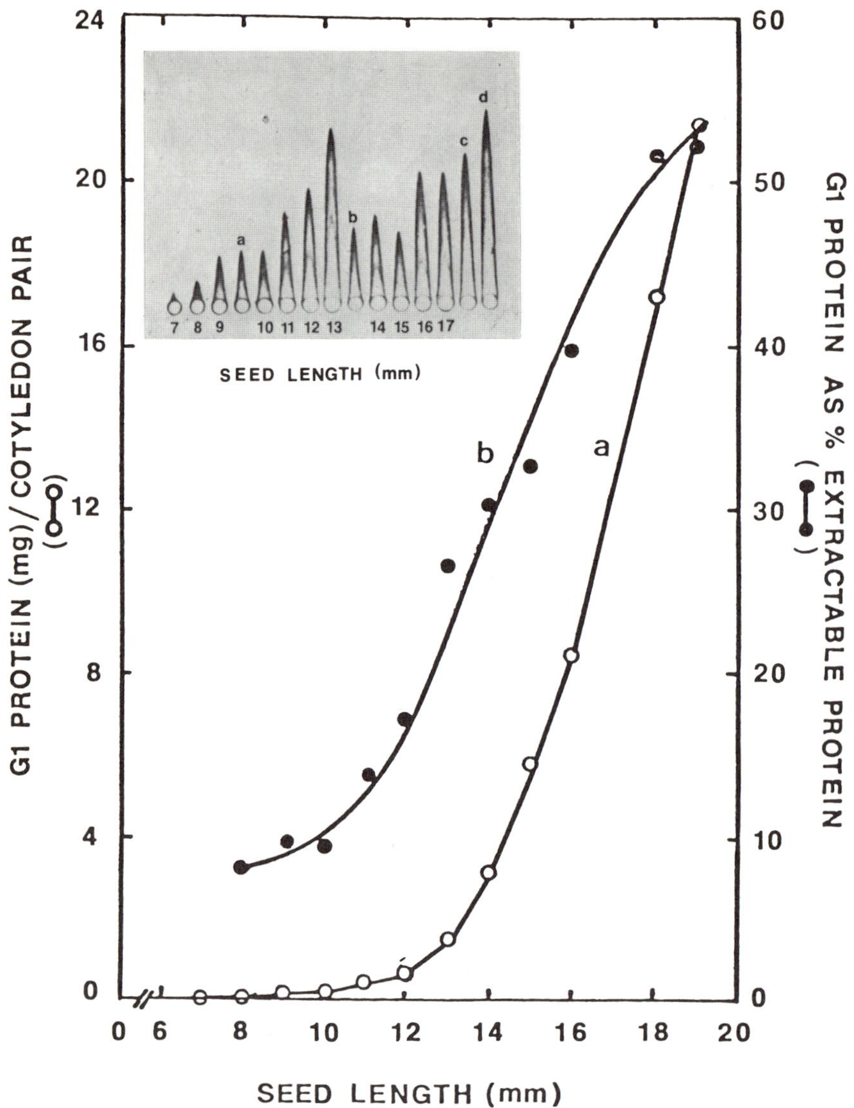

FIGURE 6. G1 protein content of developing cotyledons (○—○) determined by rocket immunoelectrophoretic techniques. The inset shows a typical rocket gel containing known quantities of G1 protein (a-d) and total protein extracts from seeds of different lengths; only the G1 protein in these extracts is immunoprecipitated. G1 protein content is also shown as a percentage of extractable protein (●—●) (from Sun et al., *1978; courtesy of The American Society of Plant Physiologists).*

on SDS polyacrylamide gels (Fig. 5). An interesting observation is that of Sussex (personal communication), who found that abscisic acid can greatly stimulate the amount of G1 protein in young embryos. Very active synthesis and accumulation of protein takes place in the cotyledons of 13-21 mm seeds (18-40 DAF) (Fig. 7). This observation is in close agreement with the development of the ordered endoplasmic reticulum seen cytologically (Fig. 4). It is over this period that major increases in wet and dry seed weight occur (Fig. 7). From a practical point of view, this time is critical to the productivity of the crop. The sudden onset and intensive synthesis of a few protein species is also favorable for biochemical studies on protein biosynthesis, accumulation, and associated regulatory processes. An understanding of these events at the molecular level may suggest ways to enhance seed metabolism over this period and consequently to increase yields.

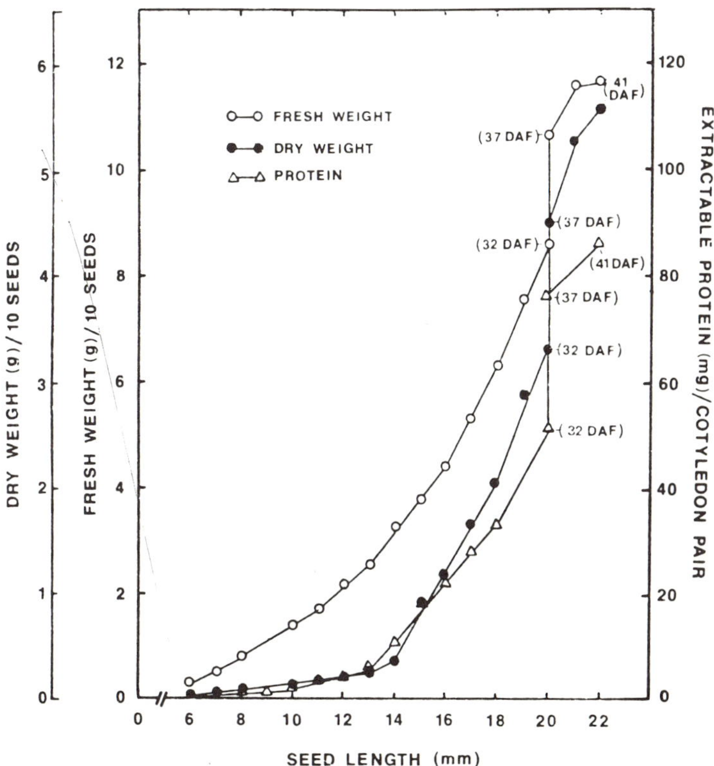

FIGURE 7. Changes in protein content, fresh and dry weight of ripening bean seeds. Although seed length corresponds well with days after flowering (DAF) up to 19 mm (see Fig. 6), there is little increase in length after 32 DAF.

STUDIES ON SEED PROTEIN BIOSYNTHESIS *IN VITRO*

The active accumulation of protein revealed by cytological, chemical, and electrophoretic analyses strongly indicated that cotyledons from bean seeds 12 to 19 mm in length should represent an excellent source for polysomes engaged in G1 protein synthesis. This was confirmed when polysomes isolated by techniques developed by Davies *et al.*, (1972) and by Verma *et al.*, (1974) were added to a cell-free protein synthesis system from wheat germ (Sun *et al.*, 1975). The protein-synthesizing activity of these polysomes and the increasing proportion of polysomes engaged in G1 synthesis as the cotyledons ripen can be judged from Fig. 8A. Immunoprecipitation of polypeptides synthesized in the polysome-directed reactions with anti-G1 serum yielded material that comigrated with each of the subunits of G1 protein (Fig. 8B), although only a relatively small amount of α subunit was detected.

ISOLATION OF G1 mRNA AND CHARACTERIZATION OF PRODUCTS TRANSLATED *IN VITRO*

The large amount of G1 protein synthesized by the growing bean cotyledons was good evidence that a relatively high proportion of G1 mRNA was present. This situation encouraged us to attempt the isolation of mRNA by a direct procedure that does not require the preparation of polysomes. Cotyledons from 12 to 19 mm long seeds were excised, frozen in liquid N_2, and then stored at -80°C. Immediately prior to extraction, cotyledons were thawed to about 0°C, then added to buffer (0.2 *M* $Na_2B_4O_7$, 1% SDS, 30 m*M* EGTA, and 5 m*M* dithiothreitol, adjusted to pH 9.0 with NaOH) which was at 100°C. About 5 g cotyledons were added per 13 ml buffer and very rapidly ground with a Polytron homogenizer, the temperature of the extract thus being about 70°C so that contaminating ribonuclease was inhibited by the hot SDS-containing buffer. The extract was deproteinized with phenol or, preferably, with proteinase K (0.5 mg enzyme/ml of buffer) by incubation for 1 hr at 37°C. After the addition of 1 ml 2 *M* KCl, the mixture was cooled to 4°C, centrifuged, and the supernatant made to 2 *M* in LiCl. Precipitation of the RNA was accelerated by briefly freezing the mixture, which was then stored at 4°C overnight, and the precipate collected by centrifugation and washed twice with 2 *M* LiCl. After dissolving in water, the precipitate was made 7% in K acetate and then reprecipitated with ethanol. Poly(A)-containing RNA was obtained by oligo(dT)-cellulose chromatography (Aviv and Leder, 1972). To eliminate contamination by ribosomal RNA, it was

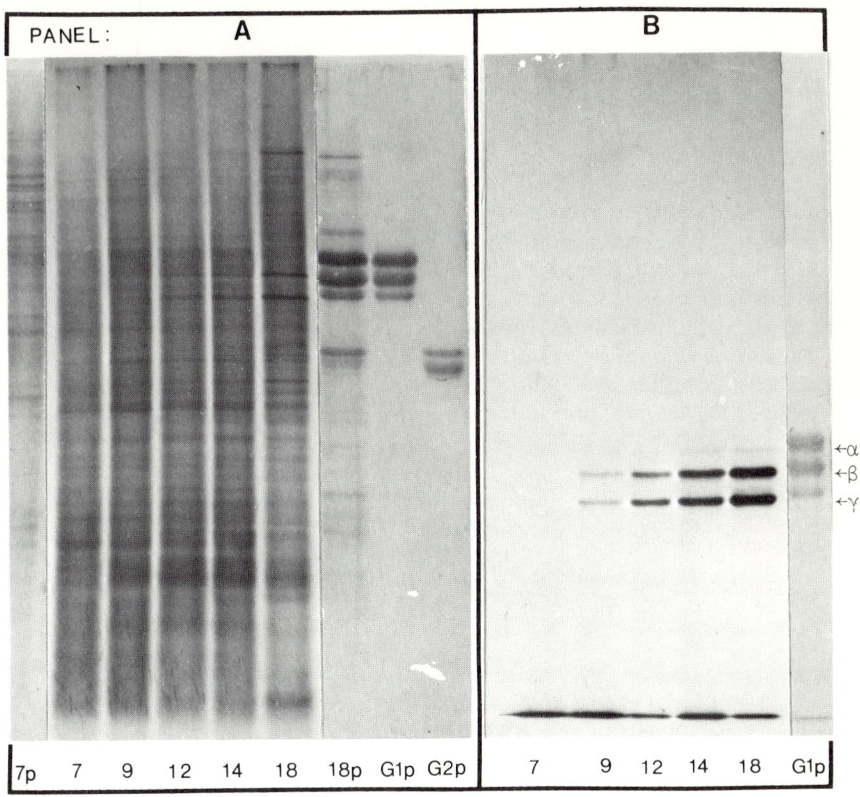

FIGURE 8. Electrophoretic analyses of products synthesized by bean cotyledon polysomes in vitro. *The numbers below each lane indicate the seed length (mm) from which polysomes were extracted. Lanes marked with a suffix p were visualized by Coomassie brilliant blue staining (G1 and G2 were authentic reference samples); all the other lanes are autoradiographs of [^{35}S]met-labeled products. For each panel, approximately equal amounts of radioactive product were added to each lane. Panel A (16% gel) shows a comparison of products directed by total polysomes isolated from seeds of different ages. Panel B (10% gel) shows the changing proportion of G1 polypeptides that can be immunoprecipitated from the total polysome-directed products as the cotyledons ripen (from Sun* et al., *1978; courtesy of The American Society of Plant Physiologists).*

found essential to heat and quench-cool the RNA immediately before application to the oligo(dT)-cellulose column. Typically, at least two cycles of purification on this column are needed. Separation of different mRNA fractions from the total poly(A)-

containing RNA was achieved by centrifugation through 0-30% linear-log sucrose gradients (Brakke and Van Pelt, 1970) as shown in Fig. 9.

The upper panels of Fig. 9 show the sedimentation profiles of RNA extracted using borate buffer followed by deproteinization by phenol (BP) or proteinase K (BK) treatment. RNA obtained by precipitation with ethanol from these gradients was used as messenger in a wheat germ cell-free protein synthesis system. The lower panels of Fig. 9 show that polypeptides of differing sizes were translated using these RNA fractions as messengers. The largest polypeptides seen in products directed by RNA from the 10S peak were of 15,000 and 18,000 daltons; from the 13S peak they were of 24,000 and 29,000 daltons. Among the products of the 16S peak were polypeptides of 47,000 and 43,000 daltons that coelectrophoresed with the β and γ sub-

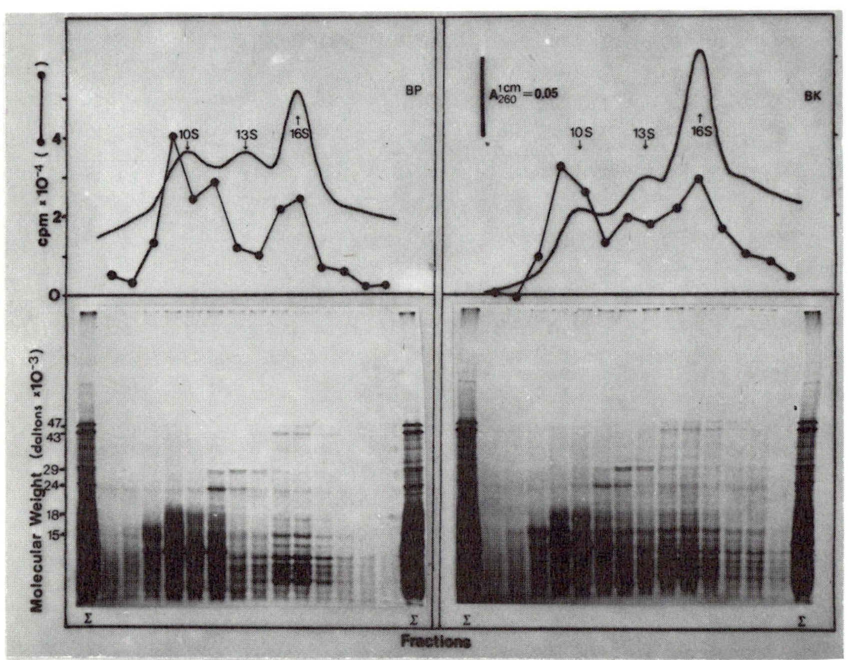

FIGURE 9. Profiles of cell-free products directed by sucrose gradient fractions. Poly(A) RNA from extracts BP and BK was separated on sucrose gradients and messenger activity assayed (top panels: Activity is for 5 μl samples of the 25 μl reactions). Polypeptide profiles of products (10 μl of the 25 μl reactions) coded by the individual RNA fractions are shown in the lower panels. Σ denotes products products coded by the unfractionated poly(A) RNAs.

units of G1 protein. About 100 μg mRNA were recovered from 5 g cotyledons; this value compares favorably with the yield obtained from polysomes, and the proportion of 47,000 and 43,000 mol. wt. polypeptides in the translation products was much higher than when polysome-derived poly(A)+RNA was used as messenger (Hall *et al.*, 1977a; 1977c).

A second cycle of purification of the 16S poly(A)+RNA by sucrose density gradient centrifugation yielded mRNA that was very efficient in directing the synthesis of 47,000 and 43,000 dalton polypeptides that were precipitable with antiserum to native G1 protein (Fig. 10A), and hence were almost certainly the β and γ subunits of G1 protein. The addition of 7-methylguanosine monophosphate greatly inhibits the translation of the 16S RNA (Hall *et al.*, 1978), strongly suggesting that the G1 mRNAs have a 7-methylguanosine cap structure (Hickey *et al.*, 1976). Further proof that the polypeptides synthesized *in vitro* are the β and γ subunits of G1 protein has been obtained (Hall *et al.*, 1978) by peptide mapping (Fig. 11), using the electrophoretic technique of Cleveland *et al.* (1977).

Variation in reaction conditions has been found to alter the proportion of 47,000 and 43,000 dalton polypeptides in the cell-free translation products, as has the use of different batches of wheat germ. However, the major difference appears to result from the quality of the 16S RNA; some preparations are capable of directing the incorporation of essentially all of the radioactive amino acid substrate into 47,000 and 43,000 mol. wt. products (Fig. 10B). From the proportion of β and γ subunits in the translation products, we estimate that in good preparations at least 50% of the 16S RNA obtained by the above extraction procedure from 12-19 mm bean cotyledons is G1 mRNA.

The absence of the α (53,000 mol. wt.) subunit of G1 protein from the translation products of the 16S RNA is puzzling. The α subunit is clearly seen among products directed by bean polysomes in the wheat germ system (see Fig. 8B) but has not been identified among translation products of free RNA at any stage of purification from total RNA to the 16S poly(A)-containing RNA fraction. We have some evidence that higher levels of Mg^{++} (8 m*M* in place of 4 m*M*) in reactions using bean polysomes enhance the amount of α subunit in the products, but variation of Mg^{++} or other conditions has not resulted in the translation of any α subunit in reactions coded by free mRNA preparations. Since each of the native G1 subunits appears to be glycosylated to about the same extent (Hall *et al.*, 1977b), it seems unlikely that the α subunit is derived from the β subunit by additional glycosylation reactions. However, peptide mapping of the individual G1 subunits is needed to provide a definitive answer to this possibility. Since the presence of a poly(A) tract is the basis for mRNA purification by oligo(dT)-cellulose chromatography, a nonpolyadenylated mRNA would be lost

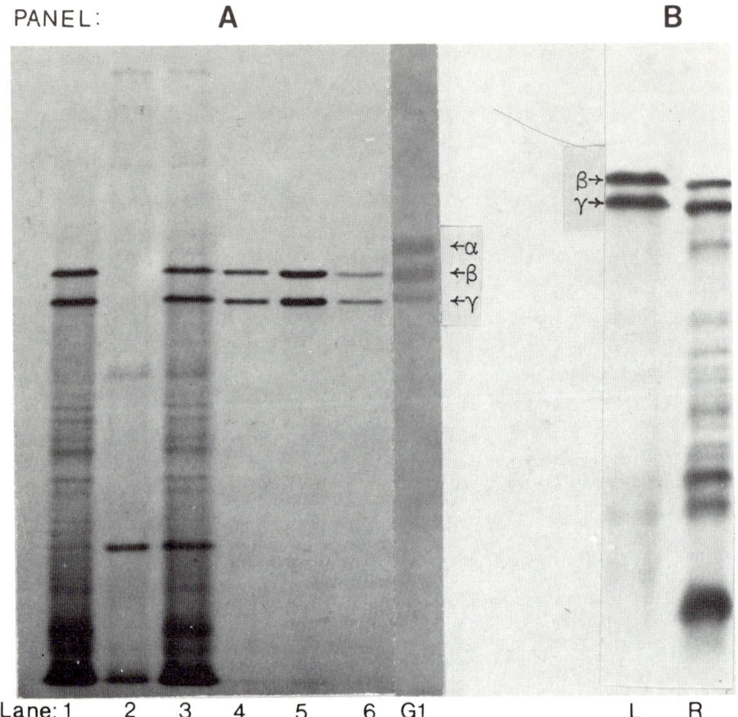

FIGURE 10. Autoradiograph of an electrophoretic separation of polypeptides synthesized in vitro. Panel A: Lane 1 contained 2.5 μl of the total polypeptide products of a 50 μl cell-free reaction coded by 0.5 μg 16S RNA from bean cotyledons. Lane 2 shows 2.5 μl of the products coded by 1 μg brome mosaic virus RNA, and lane 3 shows 5 μl of a mixture of the products shown in lanes 1 and 2. Selective immunoprecipitation of the β and γ subunits of G1 protein from products of reactions shown in lane 1 yielded material giving the electrophoretic profile seen in lane 4. Indirect immunoprecipitation using goat anti-rabbit IgG in addition to rabbit anti-G1 serum gave the profile shown in lane 5. Indirect immunoprecipitation of the mixed viral and seed mRNA-directed polypeptide products provided material seen in lane 6. Authentic G1 protein was run in the same 13% gel (anode at the bottom), and the stained polypeptide subunits are seen in lane G1. Panel B shows the different template efficiencies of two preparations of 16S RNA. The total reaction products (not immunoprecipitated) were separated on a 16% gel; almost all of the radioactive amino acid can be seen to have been incorporated into the β and γ polypeptides in the reaction products shown in Lane L, but less can be seen in those products shown in lane R (Panel A from Hall et al., 1978; courtesy of The National Academy of Sciences).

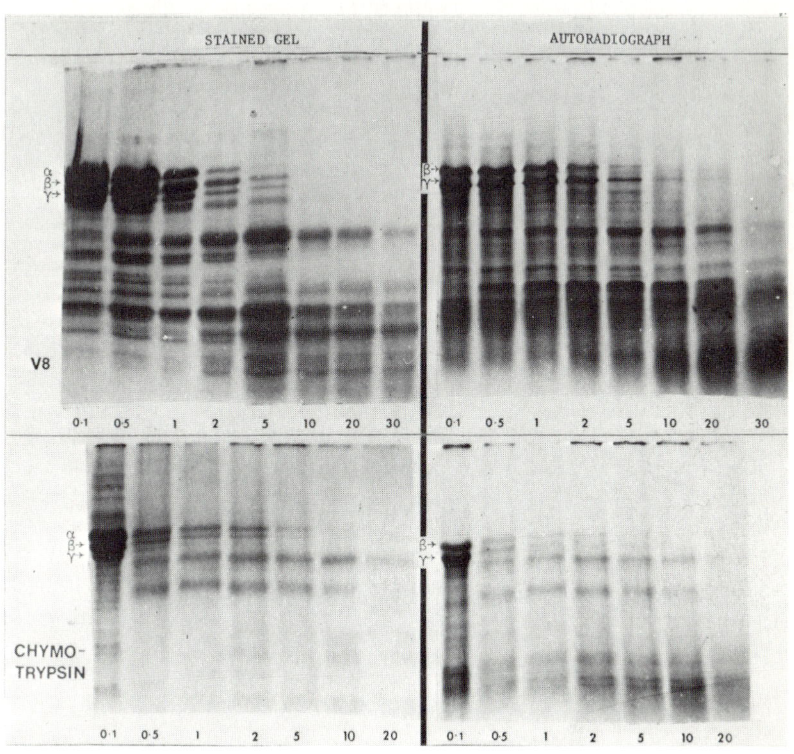

FIGURE 11. Peptide maps of authentic G1 protein and of products synthesized in vitro. Authentic G1 protein and products of cell-free synthesis directed by mRNA from developing bean cotyledons were digested with protease V8 (from Staphylococcus aureus*) or chymotrypsin. Samples were withdrawn at the times (min) indicated and subjected to electrophoresis on 20% polyacrylamide gels. For each enzyme treatment, the* Coomassie *brilliant blue-stained peptides are on the left, and autoradiograms of the [^{35}S]met-containing peptides are on the right. The similar degradation patterns for the authentic (stained) and cell-free reaction product (autoradiographed) polypeptides confirm that the material synthesized in vitro is predominantly the β and γ subunits of G1 protein (from Hall et al., 1978; courtesy of The National Academy of Sciences).*

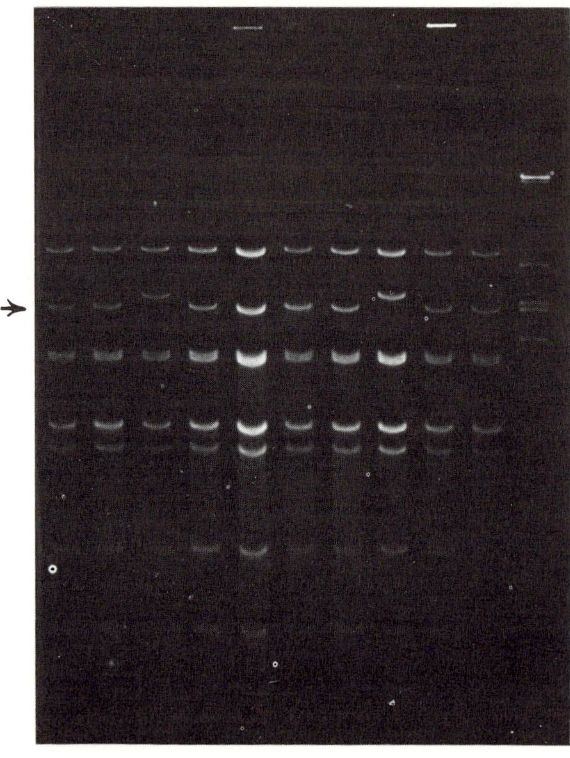

FIGURE 12. Screening of λ CH3AΔlac clones for sequences coding for G1 seed storage protein. DNA from λ CH3AΔlac clones was prepared from crude lysates (O. Smithies, unpublished) and digested with restriction endonuclease Hpa I. The resulting fragments were fractionated by agarose (1%) gel electrophoresis and the bands visualized by staining with ethidium bromide. Inserted DNA results in a change in the electrophoretic mobility of fragment 2 (arrow). Clones 40 (c) and 132 (h) have from 400-500 base pairs of additional DNA. Lanes a-j represent Hpa I digests of λ clones showing various degrees of hybridization with ^{32}P-cDNA prepared using Phaseolus *G_1 mRNA as template. Lane k is wild type λ DNA digested with EcoR1.*

at this step because it is not bound to the column. However, RNA that does not bind to the oligo(dT)-cellulose column is relatively poor at stimulating any incorporation of amino acids into hot trichloroacetic acid-insoluble material, and no polypeptides coelectrophoresing with the Gl subunits have been detected using this RNA as messenger. Possibly mRNA coding for the α subunit of Gl contains an initiation sequence that is not recognized by the wheat germ translation system, or factors present in bean cotyledons may have to be included in the protein synthesis reactions.

Synthesis of cDNA using the 16S mRNA as template has been accomplished using reverse transcriptase. In collaboration with Drs. Harvey Faber, John Kemp, and Bill Gurley (University of Wisconsin---Madison), we are cloning this cDNA in *E. coli* using Charon lambdaphage 3AΔ*lac* as a vector. Figure 12 shows that some insertion of DNA has been accomplished, but it remains to be shown unequivocally that the inserted material is part of the Gl cistron. However, these experiments represent the first step toward obtaining full-length reverse transcripts of Gl mRNA and of cloned Gl cDNA. Isotopically labeled cDNA and mRNA (iodinated *in vitro*) will be used for identification of Gl sequences obtained from bean DNA by excision with restriction endonucleases. Once an appropriate vector (such as crown gall plasmid or a DNA virus infecting plants) has been developed, the restriction fragments and cDNA will be used in attempts to transfer specific Gl genes from one bean cultivar to another and, subsequently, to different plant species.

ACKNOWLEDGMENTS

Reverse transcriptase was obtained from Life Sciences, Inc. through the Office of Program Resources and Logistics, Viral Cancer Program, Viral Onocology, Division of Cancer Cause and Prevention, National Cancer Institute, Bethesda, Maryland 20014. We are grateful to General Mills, Inc. for a generous supply of wheat germ and to Dr. Eldon Newcomb for the use of his electron microscope.

REFERENCES

Aviv, H., and Leder, P. (1972). Purification of biologically active globin messenger RNA by chromatography of oligothymidylic acid-cellulose. *Proc. Nat. Acad. Sci. (U.S.) 69,* 1408-1412.

Brakke, M. K., and Van Pelt, N. (1970). Linear-log sucrose gradients for estimating sedimentation coefficients of plant viruses and nucleic acids. *Anal. Biochem. 38,* 56-64.

Burr, B., and Burr, F. A. (1976). Zein synthesis in maize endosperm by polyribosomes attached to protein bodies. *Proc. Nat. Acad. Sci. (U.S.) 73,* 515-519.

Carr, D. J., and Skene, K. G. M. (1961). Diauxic growth curves of seeds, with special reference to French beans *(Phaseolus vulgaris* L). *Aust. J. Biol. Sci. 14,* 1-12.

Cleveland, D. W., Fischer, S. G., Kirschner, M. W., and Laemmli, U. L. (1977). Peptide mapping by limited proteolysis in sodium dodecyl sulfate and analysis by gel electrophoresis. *J. Biol. Chem. 252,* 1102-1106.

Danielsson, C. E. (1949). Seed globulins of the gramineae and leguminosae. *Biochem. J. 44,* 387-400.

Davies, E., Larkins, B. A., and Knight, R. H. (1972). Polyribosomes from peas. An improved method for their isolation in the absence of ribonuclease inhibitors. *Plant Physiol. 50,* 581-584.

Derbyshire, E., Wright, D. J., and Boulter, D. (1976). Legumin and vicilin, storage proteins of legume seeds. *Phytochem. 15,* 3-24.

Hall, T. C., Bliss, F. A., Ryan, D. S., and Sun, S. M. (1977a). The subunit structure and cell-free synthesis of the major storage protein from bean (*Phaseolus vulgaris* L) seeds. *Colloq. Int. Centre Nat. Rsch. Sci. 261,* 335-343.

Hall, T. C., McLeester, R. C., and Bliss, F. A. (1977b). Equal expression of the maternal and paternal loci for the polypeptide subunits of the major storage protein of the bean, *Phaseolus vulgaris* L. *Plant Physiol. 57,* 1122-1124.

Hall, T. C., Sun, S. M., Buchbinder, B. U., and Belozerskii, M. A. (1977c). The translation of mRNA for storage globulin of the bean, *Phaseolus vulgaris*. *In* "Translation of Natural and Synthetic Polynucleotides" (A. Legocki, ed.), pp. 217-223. University of Poznan, Poland.

Hall, T. C., Ma, Y., Buchbinder, B. U., Pyne, J. W., Sun, S. M., and Bliss, F. A. (1978). Messenger RNA for Gl protein of French bean seeds: Cell-free translation and product characterization. *Proc. Nat. Acad. Sci. (U.S.) 75,* 3196-3200.

Hickey, E. D., Weber, L. A., and Baglioni, C. (1976). Inhibition of protein synthesis by 7-methylguanosine-5'monophosphate. *Proc. Nat. Acad. Sci. (U.S.) 73*, 19-23.

Laurell, C. B. (1967). Quantitative estimation of proteins by electrophoresis in antibody-containing agarose gel. *In* "Proteins in Biological Fluids," Vol. 14 (H. Peeters, ed.), pp. 499-502. Elsevier, Amsterdam.

Ma, Y., and Bliss, F. A. (1978). Seed proteins of common bean (*Phaseolus vulgaris* L.). *Crop Sci. 18*, 431-437.

McLeester, R. C., Hall, T. C., Sun, S. M., and Bliss, F. A. (1973). Comparison of globulin proteins from *Phaseolus vulgaris* with those from *Vicia faba*. *Phytochem. 12*, 85-93.

Öpik, H. (1968). Development of cotyledon cell structure in ripening *Phaseolus vulgaris* seeds. *J. Exp. Bot. 19*, 64-76.

Osborne, T. B., and Campbell, G. F. (1898). Proteids of the pea. *J. Am. Chem. Soc. 20*, 348-362.

Pusztai, A., and Watt, W. B. (1970). Glycoprotein II: The isolation and characterization of a major antigenic and non-haemagglutinating glycoprotein from *Phaseolus vulgaris*. *Biochem. Biophys. Acta 207*, 413-431.

Romero, J., Sun, S. M., McLeester, R. C., Bliss, F. A., and Hall, T. C. (1975). Heritable variation in a polypeptide subunit of the major storage protein of the bean *Phaseolus vulgaris* L. *Plant Physiol. 56*, 776-779.

Sanders, J. H., and Alvarez, C. P. (1978). Bean production trends in Latin America. *C.I.A.T. Publication*.

Stockman, D. R., Hall, T. C., and Ryan, D. S. (1976). Affinity chromatography of the major seed protein of the bean (*Phaseolus vulgaris* L.). *Plant Physiol. 58*, 272-275.

Sun, S. M., and Hall, T. C. (1975). Solubility characteristics of globulins from *Phaseolus* seeds in regard to their isolation and characterization. *J. Agr. Food Chem. 23*, 184-189.

Sun, S. M., McLeester, R. C., Bliss, F. A., and Hall, T. C. (1974). Reversible and irreversible dissociation of globulins from *Phaseolus vulgaris* seed. *J. Biol. Chem. 249*, 2118-2121.

Sun, S. M., Buchbinder, B. U., and Hall, T. C. (1975). Cell-free synthesis of the major storage protein of the bean, *Phaseolus vulgaris* L. *Plant Physiol. 56*, 780-785.

Sun, S. M., Mutschler, M. A., Bliss, F. A., and Hall, T. C. (1978). Protein synthesis and accumulation in bean cotyledons during growth. *Plant Physiol. 61*, 918-923.

Verma, D. P. S., Nash, D. T., and Schulman, H. M. (1974). Isolation and *in vitro* translation of soybean leghemoglobin mNRA. *Nature 251*, 74-77.

Walbot, V., Clutter, M., and Sussex, I. M. (1972). Reproductive development and embryogeny in *Phaseolus*. *Phytomorphology 22*, 59-68.

Weeke, B. (1973). Rocket immunoelectrophoresis. *Scand. J. Immunol. 2*, 37-46.

MOLECULAR BASIS OF ZEIN PROTEIN SYNTHESIS IN MAIZE ENDOSPERM[1]

F. A. Burr
B. Burr

Biology Department
Brookhaven National Laboratory
Upton, New York

Zein, the major storage protein of maize, has numerous important commercial uses (McKinney, 1958). It may comprise about 50% of the seed proteins. All cereal grains contain proteins with characteristics similar to zein; generically they are called "prolamines." Their name derives from the fact that they are all rich in the amino acid proline and in amide nitrogen. Zein, which is easily obtained in substantial quantities in nearly pure form by extracting the meal with 60-70% aqueous alcohol, has been known to chemists for over 150 years. Despite this, it is still poorly defined biochemically. This paradox is probably due to its extreme insolubility in the usually employed aqueous solvents without the addition of urea or detergents or adjustment to pH extremes. Nevertheless, there are a number of organic solvents that solubilize it (Swallen and Danehy, 1947; Rees and Singer, 1956). The hydrophobicity is a reflection of its unusual amino acid composition. A number of single hydrolysis time-point experiments (Swallen and Danehy, 1947; Waldschmidt-Leitz and Metzner, 1962; Wall, 1964, Mossé, 1966; Jiménez, 1966; Sodek and Wilson, 1970) indicate that there is better than 10% proline; a high proportion of the hydrophobic amino acids leucine, alanine, tyrosine, and phenylalanine; and very few basic amino acids. Also, about 85% of the glutamyl and aspartyl carboxyl groups are in the amide form, i.e., glutamine and asparagine. It is

[1]*Research was performed at Brookhaven National Laboratory under the auspices of the U. S. Department of Energy and was supported in part by Grant GM 24057 from the National Institutes of Health.*

ISBN 0-12-602050-7

generally assumed that there is no lysine or tryptophan in zein. In solution, zein has a globular conformation and a helical content similar to more conventional proteins (Danzer *et al.*, 1975). Probably because of poor solubility, a tendency toward aggregation, partial deamidation, and contamination by other proteins, zein has been reported to be very heterogeneous with respect to molecular weight (Turner *et al.*, 1965; Mossé, 1966) and net charge (Scallet, 1947; Foster *et al.*, 1950; Turner *et al.*, 1965; Mossé, 1966; Sodek and Wilson, 1970). However, on SDS-polyacrylamide gels Misra *et al.* (1975) found that there were only two bands. Most recent estimates give the components molecular weights of approximately 19,000 and 22,500 (Burr and Burr, 1976; Lee *et al.*, 1976). These two components do not correspond to α and β zein (McKinney, 1958), which are separated on the basis of solubility (probably reflecting changes in secondary or tertiary structure and possibly covalent modification). Additional minor bands are sometimes reported on SDS-PAGE, but these seem to be dependent on the method of preparation.

Duvick (1961) proposed that zein was sequestered within protein bodies in the endosperm. This hypothesis has been borne out by the subsequent work of Wolf *et al.* (1969), Christensen *et al.*, 1968), and Burr and Burr (1976). The protein bodies are largely concentrated in the horny endosperm and are largest and most prolific in the cells internal to the subaleurone (Figs. 1 and 2) and decrease in size and number toward the center of the grain. The restricted distribution of zein protein bodies is an important point--the endosperm of maize is not a homogeneous tissue but is composed of a variety of cell types. These other tissues also contain rough endoplasmic reticulum which does not participate in zein synthesis. Hence an isolation scheme was devised to separate the zein-synthesizing endoplasmic reticulum (ER) membranes from ER cisternae with other functions.

Zein protein bodies are limited by a single membrane and have a homogeneous inner matrix (Fig. 2) which can be entirely dissolved by 70% aqueous ethanol (Fig. 3). The presence of polysomes on the surface of the limiting membrane is a striking characteristic that is not true of other types of protein bodies. In soybean, for example, the protein bodies of the cotyledons do not have polysomes associated with their membrane surface (see F. A. Burr in Beachy, this volume). The protein bodies of soybean and other legumes contain soluble proteins and are "vacuolar" in nature, whereas the protein bodies of maize store an insoluble protein and probably are derived from the endoplasmic reticulum (Khoo and Wolf, 1970). The details of differentiation are not known, but two possible schemes of development are shown in Fig. 4. Both proposed modes for the genesis of protein bodies imply that zein mRNA is translated

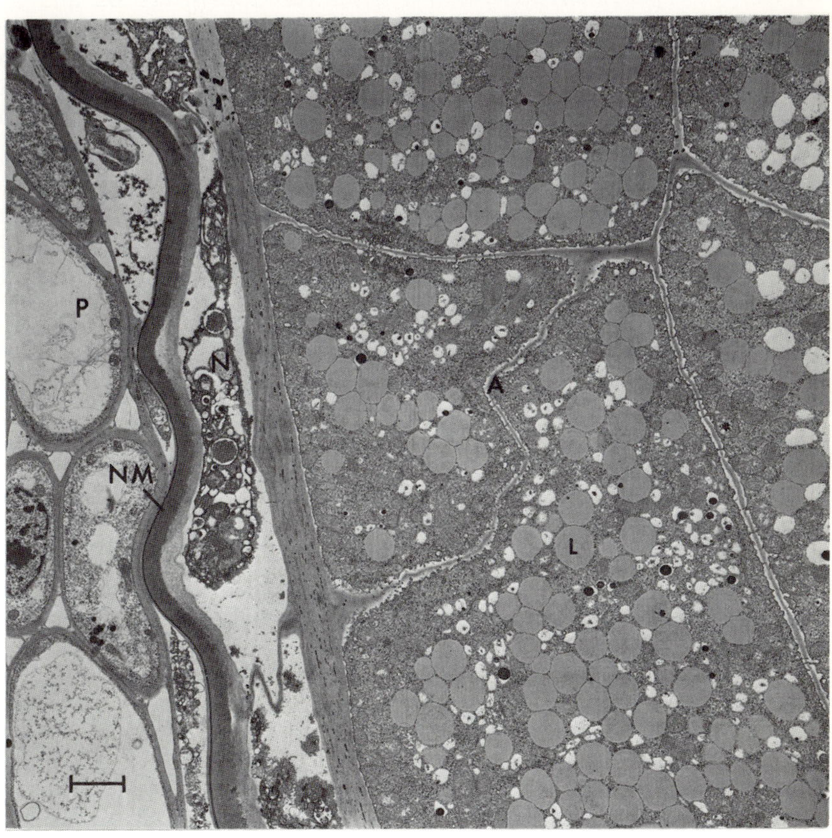

FIGURE 1. Portion of an anticlinal section through the pericarp and adjacent endosperm of the maize caryopsis 22 days postpollination: pericarp (P), nucellar membrane (NM), degenerating nucellus (N), aleurone (A). The predominant organelle in the aleurone and subaleurone is the lipid globule (L). There are also plastids, mitochondria, vacuoles (electron transparent spheres, some with dark globular or granular inclusions), microbodies (gave a positive test for catalase), and rough endoplasmic reticulum. These latter, however, do not give rise to protein bodies in these tissues (Bar = 3 μm).

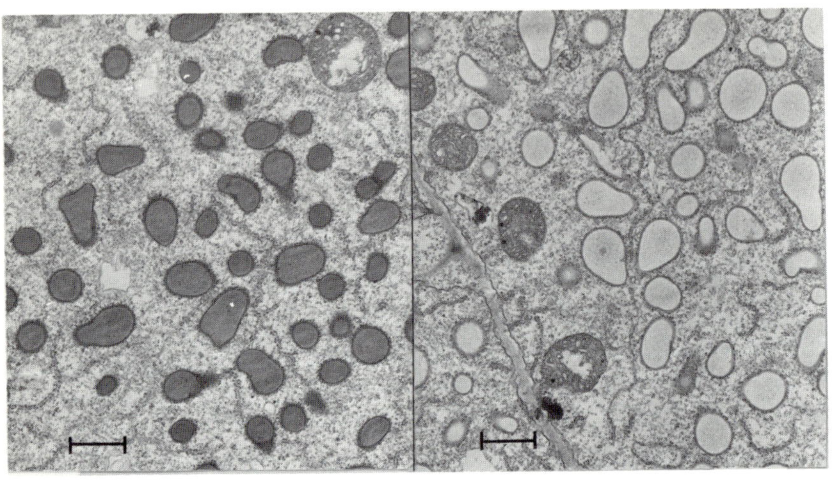

FIGURE 2. Portion of a cell in the horny endosperm 4-5 cells below the aleurone, 22 days postpollination, showing the distribution and general features of the zein protein bodies and rough endoplasmic reticulum (Bar = 1 μm).

FIGURE 3. Part of a cell similar to that shown in Fig. 2 which was extracted with 70% ethanol after glutaraldehyde fixation and prior to postosmication. The zein which constitutes the matrix of the protein bodies is completely removed by this procedure (Bar = 1 μm).

on the protein body membrane and attached endoplasmic reticulum cisternae. At present, the second pathway involving the progressive enlargement of the original cisternum is favored, as no evidence of blebbing off has been observed. If, however, budding and separation is the actual mode of development, at least continual budding can probably be discounted as Duvick (1961) noted that the protein bodies did not seem to increase in number in any given part of the endosperm after their initial appearance, but only seemed to increase in size.

A calculation can be made to see if there is enough mRNA associated with the protein bodies as they are isolated to account for the zein they contain. If the protein body develops as believed via the pathway depicted on the right side of Fig. 4, the membrane surface area and associated polysomes would remain relatively constant during protein body filling. Protein bodies isolated 15 days after pollination are a maxi-

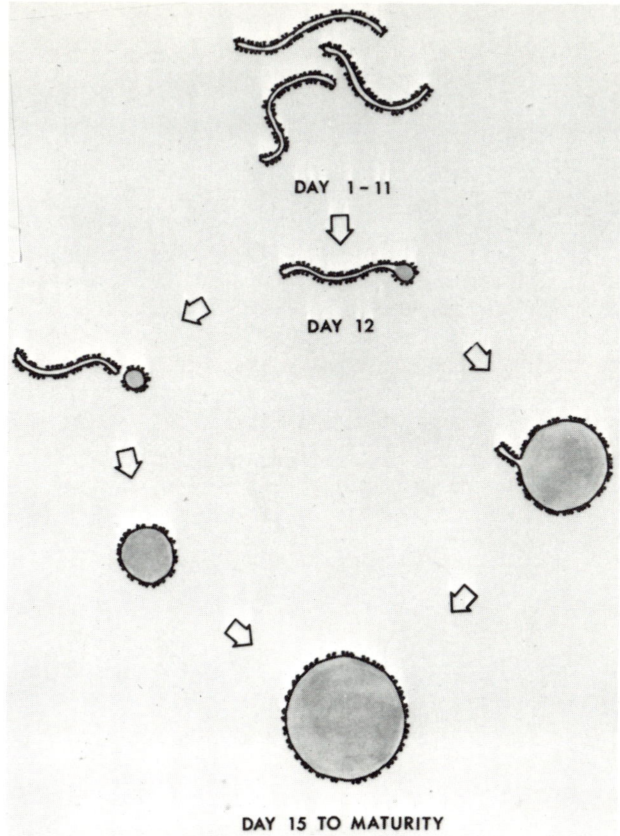

FIGURE 4. Schematic development of the protein bodies in maize. Up to day 11 postpollination there are no discernible protein bodies; there is only rough ER. About day 12 there appear peripheral swellings of the ER that contain a dark matrix. Two possibilities for the next stage in development are next depicted: Left, the initial bleb separates from the ER and progressively increases in size by additional membrane growth while zein concomitantly accumulates. Right, an alternative possibility in which the original cisternum continues to fill up (like a balloon) with no further membrane increments. By day 15 the protein body has its mature structure. The diagram is based in part on Khoo and Wolf (1970). Changes, if any, in the final maturation of the protein body remain to be defined. However, we do know that zeins extracted from immature endosperms behave similarly on SDS-polyacrylamide and urea-polyacrylamide gels.

mum of 3 days old. They contain 25 μg polysomal RNA per mg protein which is mostly zein (Burr and Burr, 1976). From our recent work on the characterization of zein mRNA (Burr *et al.*, 1978) we know that the 1.1 kilobase message constitutes 2.7% of the polysomal RNA, and we can therefore calculate that there are 2.54×10^4 molecules of zein per messenger RNA. Estimating that the average zein molecule has 188 amino acids and knowing that there are on the average 7.5 ribosomes associated with each message (Burr and Burr, 1976; Burr *et al.*, 1978), it would require an elongation rate of 2.46 amino acids/ribosome/sec to make this much zein in three days. Hemoglobin is reported by various authors to be polymerized at the rate of 1.7 to 6 amino acids/sec in intact reticulocytes (Palmiter, 1973); so this calculation for zein is not an unreasonable rate and the mRNA associated with zein protein bodies could well account for all the zein contained.

In examining the initial electron micrographs, it occurred to us that the zein might be synthesized at its site of deposition by the attached polysomes. This would obviate the difficulties of transporting such an insoluble protein through the aqueous cytoplasm. In 1976 we reported that we were able to separate the zein protein bodies from other cellular components (Burr and Burr, 1976). When zein polysomes eluted from the isolated protein bodies were placed in an *in vitro* amino-acid-incorporating system derived from corn (Mans and Novelli, 1964), where nascent polypeptide chains are completed but no new chains are initiated, we noted that on SDS-polyacrylamide gels the major *in vitro* products behaved as though they were just slightly smaller than authentic zein. In that earlier experiment (Burr and Burr, 1976), the released chains were examined midway on the linear phase of the synthesis curve. Quite a different picture emerges, however, if several time points during synthesis are examined. A time course of the incorporation of ^{3}H-leucine in the presence of polysomes derived from zein protein bodies is presented in Fig. 5a. The arrows indicate points at which a large excess of cold leucine was added to the reaction in parallel tubes to dilute the specific activity of ^{3}H-leucine. Fig. 5b shows the distribution of label when tritiated products from these reactions are electrophoresed on SDS-polyacrylamide gels together with ^{14}C-labeled authentic zein. In this experiment the two zein polypeptide chains are poorly resolved, but the major ^{14}C-peak in each case corresponds to the 19,000 dalton polypeptide. The chains completed early in the synthesis reaction are migrating considerably faster than the authentic protein. However, as time progresses and peptides from initially smaller nascent chains are completed, the *in vitro* products approach and may even exceed the size of authentic zein. We believe that native zein undergoes at least two postsynthetic proces-

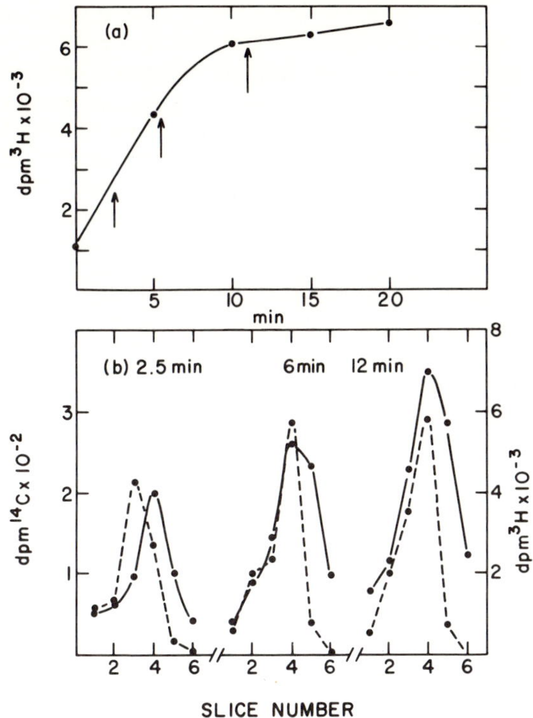

FIGURE 5. Polysome-dependent in vitro *protein synthesis. To prepare a high speed supernatant, 3 day old etiolated shoots of WF9 x B37 maize were ground in sterile, cold 20 mM HEPES, pH 7.5, 120 mM potassium acetate, 5 mM magnesium acetate, 0.45 M sucrose. A postribosomal supernatant was desalted on a Sephadex G-25 column at 4°C in the same buffer without sucrose. The pooled fractions were frozen in small aliquots and stored at -85°C. Protein synthesis reactions contained 2.55 mg/ml high-speed supernatant protein, 2.5 mM ATP, 0.375 mM GTP, 5 mM phsphoenolpyruvate, 25 μM each of 19 unlabeled amino acids, 20 mM HEPES, pH 7.6, 10.4 mM magnesium acetate, 56 mM potassium acetate, 20 μM ^{3}H-L-leucine (330 mCi/m mole), 20 μg/ml pyruvate kinase, and 0.12 mg polysomal RNA per ml. Incubation was at 37°C. (a) Time course of reaction. (b) At points indicated by arrows, individual reactions were made at 10 mM in unlabeled L-leucine; at 20 min after the reactions were initiated, they were stopped with 20 mM EDTA and 20 μg/ml pancreatic ribonuclease. ^{14}C-alkylated zein was added and the reactions were extracted with 70% ethanol. The extracts were dialyzed against water, lyophilized, and electrophoresed on SDS-polyacrylamide gels as described (Burr et al., 1978). The regions of the stained gel containing zein bands were sliced into 0.5 cm sections, dissolved, and counted as described (Burr and Burr, 1976). Migration was from left to right (C^{14} label-----; H^3———).*

sing steps that may account for these observations, and we will document this more fully later. Briefly, however, zein appears to be glucosylated on the protein body membrane, and the presence of the sugar might be retarding the migration of the protein in SDS-polyacrylamide gels. Thus the *in vitro* product that was not made on a membrane and probably was not glucosylated will tend to migrate faster than the authentic protein. Second, zein mRNA translated in an initiating, messenger-dependent system produces products larger than authentic zein due to the translation of an N-terminal signal peptide. This peptide is likely cleaved from each zein chain as it passes through the protein body membrane. It is possible that when zein polysomes were eluted from the protein body membrane some of the large nascent chains had already lost their signal peptides. The chains first completed in polysome runoff are probably migrating faster than authentic zein because they have been cleaved to the same length and lack the sugar. The addition of the sugar must occur late enough in the synthesis and packaging process that very few if any polysomes were prepared containing nascent chains that had been both cleaved and glucosylated.

The observation that the products of *in vitro* synthesis in the polysome runoff system were smaller than authentic zein led us to speculate that a component might be added postsynthetically to the protein. We first thought to look for the presence of sugar on zein. Table I summarizes a number of experiments carried out in 1976 to quantitate the amount of sugar associated with zein. It can be seen that zein prepared in a variety of ways from two different strains and two maturity dates has rather consistently been found to have about 1 mole of neutral sugar per 20,000 g of protein. Tests for amino sugars have been negative. Evidence that the sugar is associated with both polypeptide chains is shown in Fig. 6. Zein has been electrophoresed on SDS-polyacrylamide gels and stained by the periodic acid-Schiff's procedure. Included in these gels were a positive control, ovalbumin, a glycoprotein, and a negative control, cytochrome c, a nonglycoprotein.

In order to identify the neutral sugar, deionized hydrolysates were chromatographed on thin-layer cellulose sheets along with authentic sugars. Fig. 7 illustrates one such separation. Xylose was added to the zein hydrolysates to serve as an internal standard of migration behavior. (When the zein hydrolysate is chromatographed alone no spot with the migration of xylose is observed.) The hydrolysate shows only one spot with a migration corresponding to glucose. In two other sol-

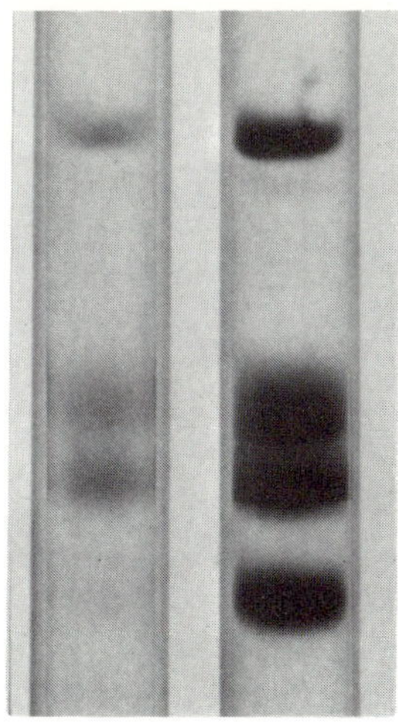

Fig. 6

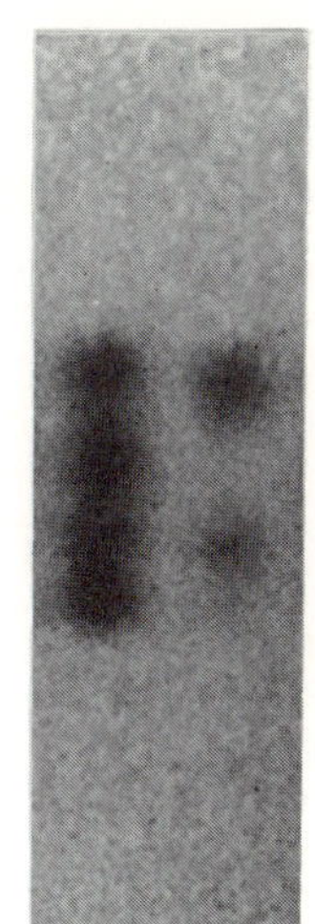

Fig. 7

FIGURE 6. 10% SDS-polyacrylamide tube gels. Bands from top to bottom in both gels are: ovalbumin (10 μg), zein 22.5k and 19k (100 μg total), cytochrome c (10 μg). Right gel has been stained for protein with Coomassie brilliant blue. Left gel has been stained for carbohydrate by the periodic acid-Schiff's reaction. Cytochrome c does not stain in the latter treatment as it is a nonglycoprotein. A large amount of zein was necessary because the quantitative data showed that the amount of sugar was very low--1 mole sugar/mole polypeptide chain.

FIGURE 7. Chromatography on Eastman Kodak thin layer cellulose. Solvent system = n-butanol:pyridine:water (Vomhof and Tucker, 1965). Spray detection = ammoniacal silver nitrate (Trevelyan et al.*, 1950). Left track (from top to bottom), sugar standards: xylose, mannose, glucose, galactose; right track, zein neutal sugar preparation containing xylose as an internal marker.*

vent systems--ethyl acetate:pyridine:water (Schweiger, 1962) and n-butanol:acetone:water (Ghebregzabher *et al.*, 1976) the major spot comigrates with glucose. However, in these two latter systems there is also some slower migrating material present. We do not know if this indicates the presence of another component or an alteration of glucose during hydrolysis.

TABLE I. Quantitation of Neutral and Amino Sugars Associated with Zein

Expt.	Preparation method[a]	Assay method	Std.	moles/20,000 g
1	70% ethanol[b]	phenol-H_2SO_4[c]	glucose	0.7
2	70% ethanol; 1 N H_2SO_4, 4 hr, 100°C;			
	passed through Dowex 50 and Dowex 1[d];	phenol-H_2SO_4	mannose	0.72
	6 N HCl, 3 hrs, 100°C; analyzed directly	Elson-Morgan[e]	hexosamine	0.0895
3	Reduced and alkylated zein eluted from			
	SDS hydroxylapapatite;[f] 6 N HCl, 3 and			
	13 hrs, 100°C; eluted from Dowex 50[g]	Elson-Morgan	galactosamine	0.058
4	70% ethanol; 1 N H_2SO_4, 4 hrs, 100°C;			
	passed through Dowex 50 and Dowex 1	phenol-H_2SO_4	glucose	1.11

5	70% ethanol; 1 N HCl, 6 hrs, 100°C;			
	passed through Dowex 50;	anthrone[h]	glucose	0.95
	analyzed directly	"	glucose	0.88
6	22 day endosperm, 70% ethanol;			
	2 N H_2SO_4, 4 hrs, 100°C;			
	passed through Dowex 50 and Dowex 1	phenol-H_2SO_4	glucose	1.1
		glucostat[i]	glucose	0.68
7	Illinois High Protein, reduced and			
	alkylated; eluted from hydroxylapatite	phenol-H_2SO_4	glucose	1.06

[a]Except where noted zein was prepared from whole, mature grain of WF9 x B37. [b]Extracted as described (Burr and Burr, 1976). [c]Ashwell, 1966. [d]Spiro, 1966. [e]Davidson, 1966. [f]Moss and Rosenblum, 1972. [g]Boas, 1953. [h]Ashwell, 1957. [i]Worthington Biochemical Corporation.

Although the usual sugar donor in glycosylation is a nucleoside diphosphate sugar, we reasoned that if zein was a glycoprotein it should be possible to label zein while it was made *in vivo* with a radioactive neutral sugar. Zein is made in nondividing cells in the endosperm. It is possible to explant young developing endosperms into culture medium and show the incorporation of radioactively labeled amino acids into zein within 24-48 hrs after transfer. Fig. 8 shows the result of an experiment in which developing endosperms were cultured in the presence of one of two neutral sugars labeled with tritium and ^{14}C-L-proline. Zein was extracted from these explants after 24 hrs and separated on SDS polyacrylamide gels. Although galactose produced a high background, there is no apparent association with zein as marked by ^{14}C-proline (Fig. 8a). On the other hand, glucose gave good labeling of both zein chains (Fig. 8b).

To summarize our work with the sugar moiety: It appears that there is one mole of neutral sugar/mole of polypeptide chain. Based on reaction with glucose oxidase, thin-layer chromatography, and *in vivo* labeling, we tentatively identify the sugar as glucose. Since glucose would be the most prevalent contaminant in our preparations, we believe that it will be necessary to demonstrate a covalent linkage before it can be definitely concluded that zein is glucosylated. We are unaware of any other protein glycosylated with a single glucose molecule. Furthermore, it should be mentioned that gliadin, the prolamine of wheat, has been examined for sugar content and has been found not to be glycosylated (Bernardin *et al.*, 1976).

Elsewhere we have reported the purification and partial characterization of the zein mRNAs (Burr *et al.*, 1978). When this messenger preparation was translated in the wheat germ messenger-dependent protein synthesizing system, both zein polypeptides were made. However, they appeared to be larger than the authentic zein polypeptides on SDS-polyacrylamide gels. Since zein mRNAs are translated on membrane-bound ribosomes, it was not surprising that the messenger dependent products were larger. In animal and bacterial systems most proteins made on membranes for extracellular transport are initially made as larger translation products with an additional sequence at their amino terminus and are subsequently cleaved to their final size by specific proteases as they pass through the membrane (Blobel and Dobberstein, 1975). Zein resembles these secretory proteins in that after or during translation it must pass through the protein body membrane to be stored in the lumen. In order to show that the larger zein translation products are also the result of an extra N-terminal sequence, we have examined the cyanogen bromide peptides from one of the two chains. In this experiment products of a wheat

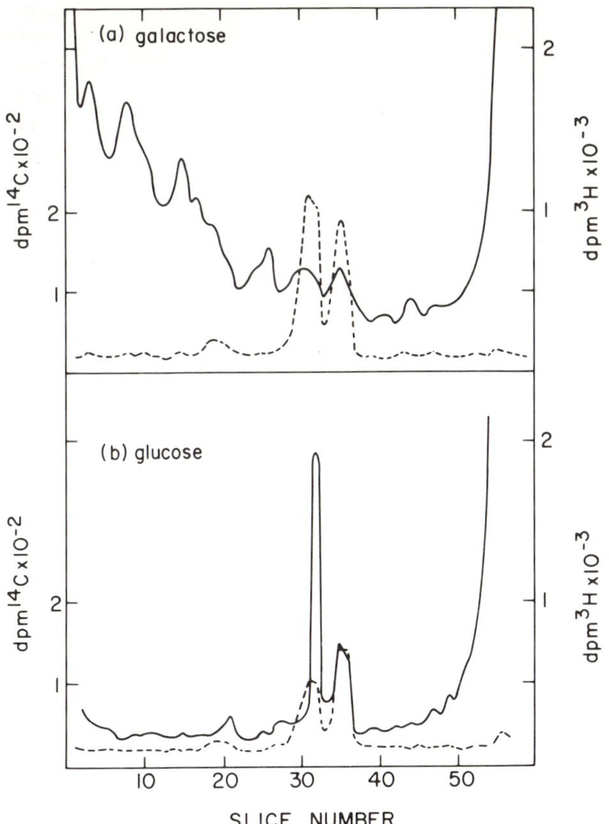

FIGURE 8. Labeling of zein in cultured endosperm explants. Illinois High Protein endosperms 15 days postpollination were dissected out of ears surface-sterilized with 25% chlorox and plated onto L broth medium overnight. The next day pieces showing no contamination were added to RMS medium (Green and Phillips, 1975) modified to contain no asparagine, 2 mg/l IAA instead of 2,4-D, and 12.5 μCi ^{14}C-proline (260 mCi/m mole)/ml. The usual amount of sucrose was replaced by 0.125 mCi D-galactose-1-^{3}H (14 Ci/m mole)ml (a) or 0.125 mCi D-glucose-1-^{3}H (18 Ci/m mole)ml (b). Tissue was incubated overnight at room temperature. Zein was extracted as described (Burr and Burr, 1976) and electrophoresed on 10% SDS-polyacrylamide tube gels. Gel slices were processed for counting as described by Burr and Burr (1976) (^{14}C label -----; ^{3}H label ———).

germ translation were acid precipitated and incubated in 70% formic acid with or without cyanogen bromide. The acid was removed by lyophilization and the products were separated by SDS-PAGE together with authentic zein which had been reduced and treated with 70% formic acid or formic acid plus cyanogen bromide. To discount minor differences in migration rates between the channels, authentic zein markers labeled with fluorescamine were included in all samples. Figure 9a shows the migrations of the authentic zein markers: on the left the 19,000 and 22,500 dalton untreated zein chains; on the right the products of cyanogen bromide treatment. It can be noted in the latter that only the upper chain cleaved and that three new bands were generated. When zein is reacted with dansyl chloride both chains are labeled. Because of the reported absence of lysine in zein we assume that only the amino termini are modified in this reaction. The conditions of reaction (Gros and Labouesse, 1969), pH 8.7-9.4, do not favor the labeling of the ε-amino group of lysine or the phenolic group of tyrosine. When dansylated zein is cleaved the only two peptides labeled are the uncleaved 19,000 dalton polypeptide and the fastest migrating or smallest peptide. We therefore believe the latter is the amino-terminal peptide of the 22,500 dalton chain. Fig. 9b shows a gel developed by fluorography. On the left the two major products of zein mRNA-dependent synthesis have migrated more slowly than authentic zein polypeptides (arrows). On the right side of Fig. 9b the products of *in vitro* synthesis subjected to cyanogen bromide treatment give a gel pattern similar to, but not identical with, authentic zein cleaved with cyanogen bromide (arrows). In this portion of the gel one sees that only the larger of the two *in vitro* products have been cleaved. Of the fragments produced by the cleavage only the largest corresponds in size to a cleavage fragment of the authentic zein. The intermediate size fragment migrates slightly faster than its corresponding authentic zein peptide. Periodic acid-Schiff's staining indicates that this is a glycopeptide; we assume that the *in vitro* fragment has not been glucosylated and therefore migrates faster. Finally, the smallest fragment migrates much slower than the authentic N-terminal peptide. This is taken as evidence that the additional molecular weight observed for the *in vitro* product is associated with the amino terminus, and it is this portion that is proteolytically cut to give the processed 22,500 dalton polypeptide chain. By analogy it is assumed, although not shown, that the additional length of the 19,000 dalton precursor is also amino terminal.

As further demonstration that the smallest cyanogen bromide peptide is the amino terminal portion of the *in vitro* product of the 22,500 dalton precursor, we performed an experiment adapted from that used by Dintzis (1961) to show that hemoglo-

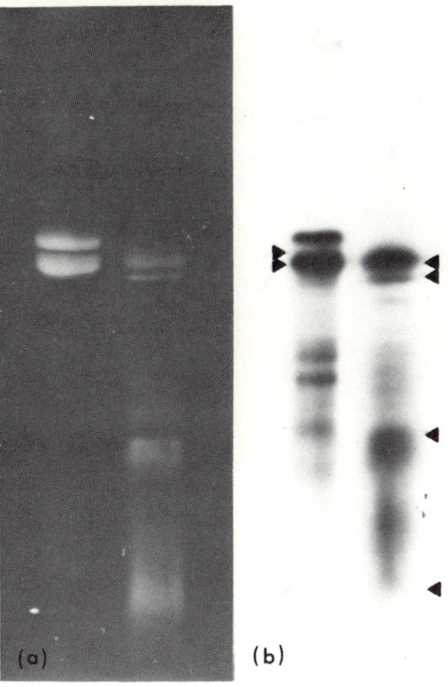

FIGURE 9. Cyanogen bromide products of authentic and in vitro *synthesized zein electrophoresed on a 20% SDS-polyacrylamide gel (Anderson* et al., *1973). (a) Zein prepared from Illinois High Protein maize as described (Burr and Burr, 1976) and reacted with or without a large excess of cyanogen bromide in 70% formic acid (16 hrs at room temperature) after reduction in 40% mercaptoacetic acid (Alfagame* et al., *1974). Following lyophilization zein was dissolved at 2 mg/ml in 4% SDS, 50 mM sodium phosphate, pH 8.2. Fluorescamine was dissolved at 3 mg/ml in N,N-dimethyl-formamide and 1/10 volume was added to the protein solutions (Knuffermann* et al., *1975). The bands were visualized with long wave UV. On the left, zein treated without cyanogen bromide; on the right, zein treated with cyanogen bromide. (b) Products of zein mRNA-dependent* in vitro *protein synthesis in the presence of ^{3}H-leucine (Burr* et al., *1978) were precipitated with 5% cold trichloroacetic acid and dissolved in 70% formic acid with or without excess cyanogen bromide and treated as above. After lyophilization the products were dissolved in sample preparation solution (Anderson* et al., *1973) and electrophoresed in the same channels as fluorescamine labeled zein. The positions of the fluorescamine labeled authentic proteins or peptides are indicated by arrows. The (Fig. 9 cont.)* in vitro *products were detected by fluorography (Bonner and Laskey, 1974). On the left, messenger products not reacted with cyanogen bromide; on the right, products exposed to cyanogen bromide.*

bin was polymerized from its amino terminal end. Zein mRNA-dependent synthesis was allowed to proceed for 15 min in the presence of ^{14}C-labeled alanine to give more or less uniform labeling throughout the polypeptide chain. Synthesis was also initiated in very high specific activity ^{3}H-leucine. At intervals aliquots were diluted with 100-fold excess of unlabeled leucine, a concentration previously determined not to affect protein synthesis, and synthesis was allowed to continue in the presence of lowered specific activity leucine. Thus at very short intervals of initial labeling, only the amino terminal portions should be labeled with high specific activity leucine, but as the labeling time increases the specific activity of the C-termini will also increase. The ^{3}H to ^{14}C ratios of the cyanogen bromide peptides (small, medium, large) and of the uncleaved 19,000 (19k) dalton precursor were measured and the data presented in Fig. 10. It can be seen that the small peptide has the highest initial $^{3}H/^{14}C$ ratio, as would be expected if it were the amino terminal peptide. The large peptide has the lowest initial $^{3}H/^{14}C$ ratio and is probably from the carboxy-terminus.

Prolamines with amino acid compositions similar to zein (Waldschmidt-Leitz and Metzner, 1962; Mossé, 1966) are found in all cereals. In most cereals they constitute an important percentage of the seed protein. This fact, coupled with the absence, or near absence, of lysine and tryptophan, accounts for the poor nutritional quality of many cereals. It was predicted that if mutants with reduced amounts of prolamine could be found, they might have a better nutritional value (Nelson, 1966). In fact, at least five such mutants have been found in maize (Mertz *et al.*, 1964; Nelson *et al.*, 1965; McWhirter, 1971; Ma and Nelson, 1975), one in barley (Ingverson *et al.*, 1973), and one in sorghum (Axtel, 1976). With the reduction in prolamine there is a concomitant reduction in the number and size of the protein bodies in opaque-2 and floury-2 maize (Wolf *et al.*, 1967) and in the barley mutant 1508 (Ingverson, 1975). In the best studied and most widely cultivated mutant, the single gene recessive opaque-2 in maize, there is 1/5 to 1/3 the normal amount of zein but a substantial increase in the salt soluble proteins (Mossé, 1966). There is no effect of the gene substitution on either the amino acid composition or the electrophoretic mobility of zein (Mossé, 1966). One way in which these widely occurring mutants could all reduce prolamine accumulation without apparently changing the protein would be if the genetic lesion involved the protein synthetic apparatus

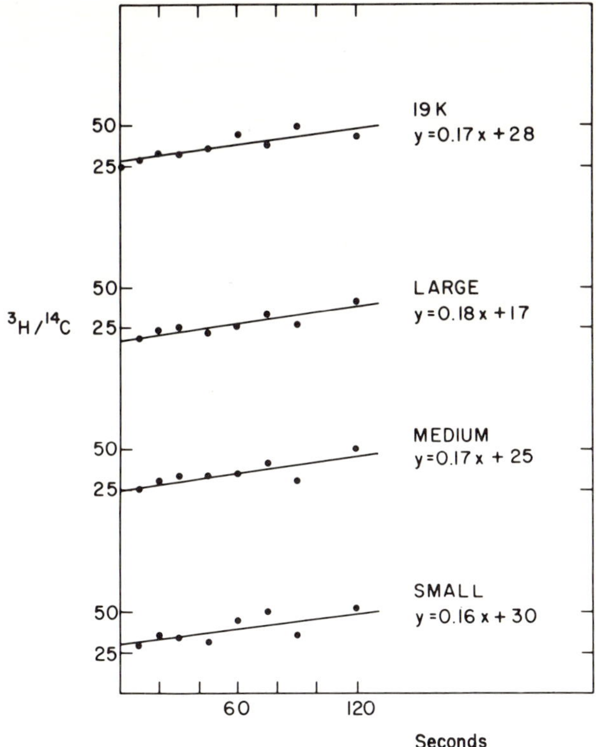

FIGURE 10. Analysis of cyanogen bromide products of in vitro *synthesis. Messenger-dependent protein synthesis as described (Burr* et al.*, 1978). Reaction was initiated with 6 μg/ml zein mRNA in the presence of 120 μM ^{14}C-alanine (135 μCi/ μmole) and 20 μM ^{3}H-leucine (46 mCi/ μmole). At the intervals indicated, 1 μl of 0.1 M L-leucine was added to a 50 μl reaction, and synthesis continued for a total of 15 min. Reactions were stopped with 5 μg pancreatic ribonuclease and 5 μl 0.2 M EDTA. Each reaction was precipitated, reacted with cyanogen bromide, and separated on SDS-polyacrylamide gels as in Fig. 7b. Bands were located by fluorography, cut from the gels, and counted (Burr and Burr, 1976). Slopes and intercepts were obtained by linear regression.*

for prolamine synthesis. Zein synthesis in maize requires translation on the protein body membrane and two postsynthetic processing steps. Zein appears to be the only protein translated at this site, which implies that specific receptors for either the message or the nascent protein may be associated with the membrane. Blocks at any stage in this synthetic process would reduce zein storage and could account for the change in the proportion of proteins observed in the mutants. Misra *et al.* (1972) observed that a number of mutants blocked in starch synthesis also had reduced levels of zein. This effect was particularly pronounced in double mutants with opaque-2. These observations could be explained by assuming that the mutants limit the substrate for glucosylation of zein. If the nascent zein chain cannot be glucosylated, it may disrupt membrane transport, slow translation, and tie up sites for synthesis.

SUMMARY

Zein, the storage protein of maize, is translated from messenger RNA on ribosomes bound to the outer membrane of the zein protein bodies. No other proteins appear to be made on this membrane. Before zein is transported through the protein body membrane it undergoes at least two posttranslational modifications: (a) Zein polypeptides made in a messenger-dependent protein synthesizing system from purified zein mRNA are larger than zein polypeptides extracted from maize seed. The newly translated polypeptides each have an additional sequence of 10-20 amino acids at their amino terminus. This "signal peptide" is probably cleaved from the nascent protein chain as it passes through the membrane. (b) Each of the zein polypeptide chains is associated with one glucose molecule. ADP-glucose is the major glucosyl donor in starch synthesis, and glucosylation of zein may be a step that regulates zein accumulation relative to saccharide biosynthesis. If zein, and only zein, is translated and processed on the protein body membrane then transported through it, it will be possible to identify factors responsible for this specificity. We believe that a number of mutations that decrease zein levels but do not alter the protein structure, can be explained in the framework of this specialized protein synthetic apparatus.

REFERENCES

Alfagame, C. R., Zweidler, A., Mahowald, A., and Cohen, L. A. (1974). Histones of *Drosophila* embryos. *J. Biol. Chem. 249,* 3729-3736.

Anderson, C. W., Baum, P. R., and Gesteland, R. F. (1973). Processing of adenovirus 2-induced proteins. *J. Virol. 12,* 241-252.

Ashwell, G. (1957). Colorometric analysis of sugars. *In* "Methods in Enzymology," Vol. 3 (S. P. Colowick and N. D. Kaplan, eds.), pp. 73-105. Academic Press, New York.

Ashwell, G. (1966). New colorometric methods of sugar analysis. *In* "Methods in Enzymology," Vol. 8 (E. F. Neufeld and V. Ginsburg, eds.), pp. 85-95. Academic Press, New York.

Axtel, J. D. (1976). Naturally occurring and induced genotypes of high lysine sorghum. *In* "Third Research Coordination Meeting of FAP/IAEA/GSF Seed Protein Improvement Program, IAEA." Vienna.

Bernardin, J. E., Saunders, R. M., and Kasarda, D. D. (1976). Absence of carbohydrate in celiac-toxic A-Gliadin. *Cereal Chem. 53,* 612-614.

Blobel, G., and Dobberstein, B. (1975). Transfer of proteins across membranes. *J. Cell Biol. 67,* 835-851.

Boas, N. F. (1953). Method for determination of hexosamines in tissues. *J. Biol. Chem. 204,* 553-563.

Bonner, W. M., and Laskey, R. A. (1974). A film detection method for tritium-labeled proteins and nucleic acids in polyacrylamide gels. *Eur. J. Biochem. 46,* 83-88.

Burr, B., and Burr, F. A. (1976). Zein synthesis in maize endosperm by polyribosomes attached to protein bodies. *Proc. Nat. Acad. Sci. (U.S.) 73,* 515-519.

Burr, B., Burr, F. A., Rubenstein, I., and Simon, M. N. (1978). Purification and translation of zein messenger RNA from maize endosperm protein bodies. *Proc. Nat. Acad. Sci. (U.S.) 75,* 696-700.

Christianson, D. D., Nielsen, H. C., Khoo, U., Wolf, M. J., and Wall, J. S. (1968). Isolation and chemical composition of protein bodies and matrix proteins in corn endosperm. *Cereal Chem. 46,* 372-381.

Danzer, L. A., Ades, H., and Rees, E. D. (1975). The helical content of zein, a water insoluble protein, in nonaqueous solvents. *Biochim. Biophys. Acta 386,* 26-31.

Davidson, E. A. (1966). Analysis of sugars found in mucopolysaccharides. *In* "Methods in Enzymology," Vol. 8 E. F. Neufeld, and V. Ginsburg, eds.), pp. 52-60. Academic Press, New York.

Dintzis, H. M. (1961). Assembly of the peptide chains of hemoglobin. *Proc. Nat. Acad. Sci. (U.S.) 47,* 247-261.

Duvick, D. N. (1961). Protein granules of maize endosperm cells. *Cereal Chem. 38,* 374-385.

Foster, J. F., Yang, J. T., and Yui, N. H. (1950). Extraction and electrophoretic analysis of the proteins of corn. *Cereal Chem. 27,* 477-487.

Ghebregzabher, M., Rufini, S., Monaldi, B., and Lato, M. (1976). Thin-layer chromatography of carbohydrates. *J. Chromatogr. 127,* 133-162.

Green, C. E., and Phillips., R. L. (1975). Plant regeneration from tissue cultures of maize. *Crop Sci. 15*, 417-421.
Gros, C., and Labouesse, B. (1969). Study of the dansylation reaction of amino acids, peptides and proteins. *Eur. J. Biochem. 7*, 463-470.
Ingverson, J. (1975). Structure and composition of protein bodies from wild type and high-lysine barley endosperm. *Hereditas 81*, 69-95.
Ingverson, J., Køie, B., and Doll, H. (1973). Induced seed protein mutant of barley. *Experientia 29*, 1151-1152.
Jiménez, J. R. (1966). Protein fractionation studies of high lysine corn. *In* "Proceedings of the High Lysine Corn Conference (E. T. Mertz, and O. E. Nelson, eds.), pp. 74-79. Corn Refiners Assoc., Inc., Washington, D. C.
Khoo, U., and Wolf, M. J. (1970). Origin and development of protein granules in maize endosperm. *Amer. J. Bot. 57*, 1042-1050.
Knufermann, H., Bhakdi, S., and Wallach, D. F. H. (1975). Rapid preparative isolation of major erythrocyte membrane proteins using polyacrylamide gel electrophoresis in sodium dodecylsulfate. *Biochim. Biophys. Acta 389*, 464-476.
Lee, K. H., Jones, R. A., Dalby, A., and Tsai, C. Y. (1976). Genetic regulation of storage protein content in maize endosperm. *Biochem. Genet. 14*, 641-650.
Ma, Y., and Nelson, O. E. (1975). Amino acid composition and storage proteins in two new high-lysine mutants of maize. *Cereal Chem. 52*, 412-419.
Mans, R. J., and Novelli, G. D. (1964). Stabilization of the maize seedling amino acid incorporating system. *Biochim. Biophys. Acta 80*, 127-136.
McKinney, L. L. (1958). Zein. *In* "Encyclopedia of Chemistry, Suppl.", G. L. Clark, ed.), pp. 319-320. Reinhold, New York.
McWhirter, K. S. (1971). A floury endosperm, high lysine locus on chromosome 10. *Maize Genet. Coop. Newsl. 45*, 184.
Mertz, E. T., Bates, L. S., and Nelson, O. E. (1964). Mutant gene that changes protein composition and increases lysine content of maize endosperms. *Science 145*, 279-280.
Misra, P. S., Jambunathan, R., Mertz, E. T., Glover, D. V., Barbosa, H., and McWhirter, K. S. (1972). Endosperm protein synthesis in maize mutants with increased lysine content. *Science 176*, 1425-1427.
Misra, P. S., Mertz, E. T., and Glover, D. V. (1975). Characteristics of proteins in single and double endosperm mutants of maize. *In* "High-Quality Protein Maize," Proceedings of the CIMMYT-Purdue Symposium on Protein Quality in Maize, pp. 291-305. Dowden, Hutchinson & Ross, Inc., Stroudsburg, Pennsylvania.

Moss, B., and Rosenblum, E. N. (1972). Hydroxylapatite chromatography of sodium dodecyl sulfate complexes. *J. Biol. Chem. 247,* 5194-5198.

Mossé, J. (1966). Alcohol-soluble proteins of cereal grains. *Fed. Proc. 25,* 1663-1669.

Nelson, O. E. (1966). Mutant genes that change the composition of maize endosperm proteins. *Fed. Proc. 25,* 1676-1678.

Nelson, O. E., Mertz, E. T., and Bates, L. S. (1965). A second mutant gene affecting the amino acid pattern of maize endosperm proteins. *Science 150,* 1469-1470.

Palmiter, R. E. (1973). Ovalbumin messenger ribonucleic acid translation. *J. Biol. Chem. 248,* 2095-2106.

Rees, E. D., and Singer, S. J. (1956). A preliminary study of the properties of proteins in some nonaqueous solvents. *Arch. Biochem. Biophys. 63,* 144-159.

Scallet, B. L. (1947). Zein solutions as association-dossociation systems. *J. Am. Chem. Soc. 69,* 1602-1608.

Schweiger, A. (1962). Trennung einfacher Zucher auf Cellulose-Schichten. *J. Chromatogr. 9,* 374-376.

Sodek, L., and Wilson, C. M. (1970). Incorporation of leucine-^{14}C and lysine-^{14}C into protein in the developing endosperm of normal and *opaque-2* corn. *Arch. Biochem. Biophys. 140,* 29-38.

Spiro, R. G. (1966). Analysis of sugars found in glycoproteins. *In* "Methods in Enzymology," Vol. 8 (E. F. Neufeld, and V. Ginsburg, eds.), pp. 3-26. Academic Press, New York.

Swallen, L. C., and Danehy, J. P. (1947). Zein. *Colloid Chem. 6,* 1140-1148.

Trevelyan, W. E., Procter, D. P., and Harrison, J. S. (1950). Detection of sugars on paper chromatograms. *Nature 166,* 444-445.

Turner, J. E., Boundy, J. A., and Dimler, R. J. (1965). Zein: A heterogeneous protein containing disulfide-linked aggregates. *Cereal Chem. 42,* 452-461.

Vomhof, D. W., and Tucker, T. C. (1965). The separation of simple sugars by cellulose thin-layer chromatography. *J. Chromatogr. 17,* 300-306.

Waldschmidt-Leitz, E., and Metzner, P. (1962). Uber die Prolamine aus Weizen, Roggen, Mais und Hirse. *Hoppe-Seyler's Zeitschrift für Physiologische Chemie 329,* 51-61.

Wall, J. S. (1964). Cereal Proteins. *In* "Proteins and their Reactions, Symposium on Foods" (H. W. Schultz, and A. F. Anglemeier, eds.), pp. 315-341. Avi Publishing Co., Westport, Connecticut.

Wolf, M. J., Khoo, U., and Seckinger, H. L. (1967). Subcellular structure of endosperm protein in high lysine and normal corn. *Science 157*, 556-557.

Wolf, M. J., Khoo, U., and Seckinger, H. L. (1969). Distribution and subcellular structure of endosperm protein in varieties of ordinary and high-lysine maize. *Cereal Chem. 46*, 253-263.

THE PLANT SEED
Development, Preservation, and Germination

THE MECHANISM OF ZEIN SYNTHESIS AND DEPOSITION IN PROTEIN BODIES OF MAIZE ENDOSPERM

Brian A. Larkins
Nina L. Pearlmutter
William J. Hurkman

Department of Botany and Plant Pathology
Purdue University
West Lafayette, Indiana

Electron micrographs of developing maize endosperm show structures common to most cereal endosperms: Starch grains, protein bodies, mitochondria, and extensive rough endoplasmic reticulum (RER) (Fig. 1). The protein bodies are surrounded by membranes, and polyribosomes can often be observed bound to the membrane surface. Different investigators studied the ultrastructure of protein bodies in developing cereal endosperms and proposed mechanisms for their formation. In an early study of wheat endosperm development, Morton and Raison (1964) noted polyribosomes attached to the membrane surrounding the protein body and suggested that the protein body formed as a highly differentiated organelle (plastid) containing polyribosomes distinct from those of the general endoplasmic reticulum (Fig. 2, structure A). In a study of maize endosperm development, Khoo and Wolf (1970) noted the occurrence of small, presumably developing, protein bodies in dilated tips of RER cisternae (Fig. 2, structure B) and suggested that protein bodies formed in vesicles produced by ER or formed at the enlarged ends of ER. Alternatively, protein bodies may form simply as nonspecific deposits within the RER. (Fig. 2, structure C).

The proposed mechanisms for protein body formation have important implications for the purification of storage protein mRNAs. If protein bodies are highly differentiated organelles, then isolation of polysomes from protein body membranes would facilitate purification of these mRNAs (Burr *et al.*, 1977). Because in our studies of zein synthesis, polysomes or mRNAs were isolated from a mixed sample of RER and protein body

ISBN 0-12-602050-7

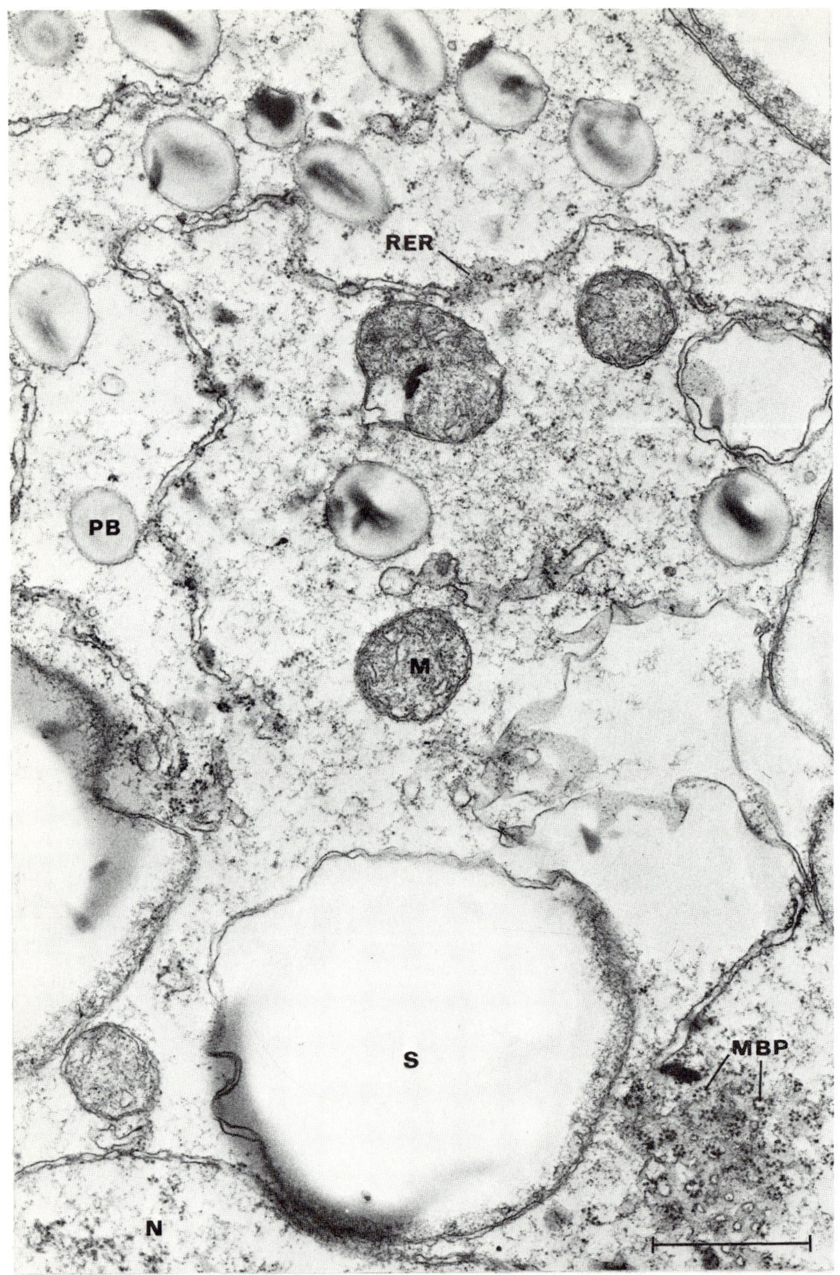

FIGURE 1. Electron micrograph of 19-day maize endosperm cell, RER, rough endoplasmic reticulum; PB, protein body; M, mitochrondrion; S, starch grain (amyloplast); MBP, membrane bound polysomes; N, nucleus (Scale: bar = 1.0 μm).

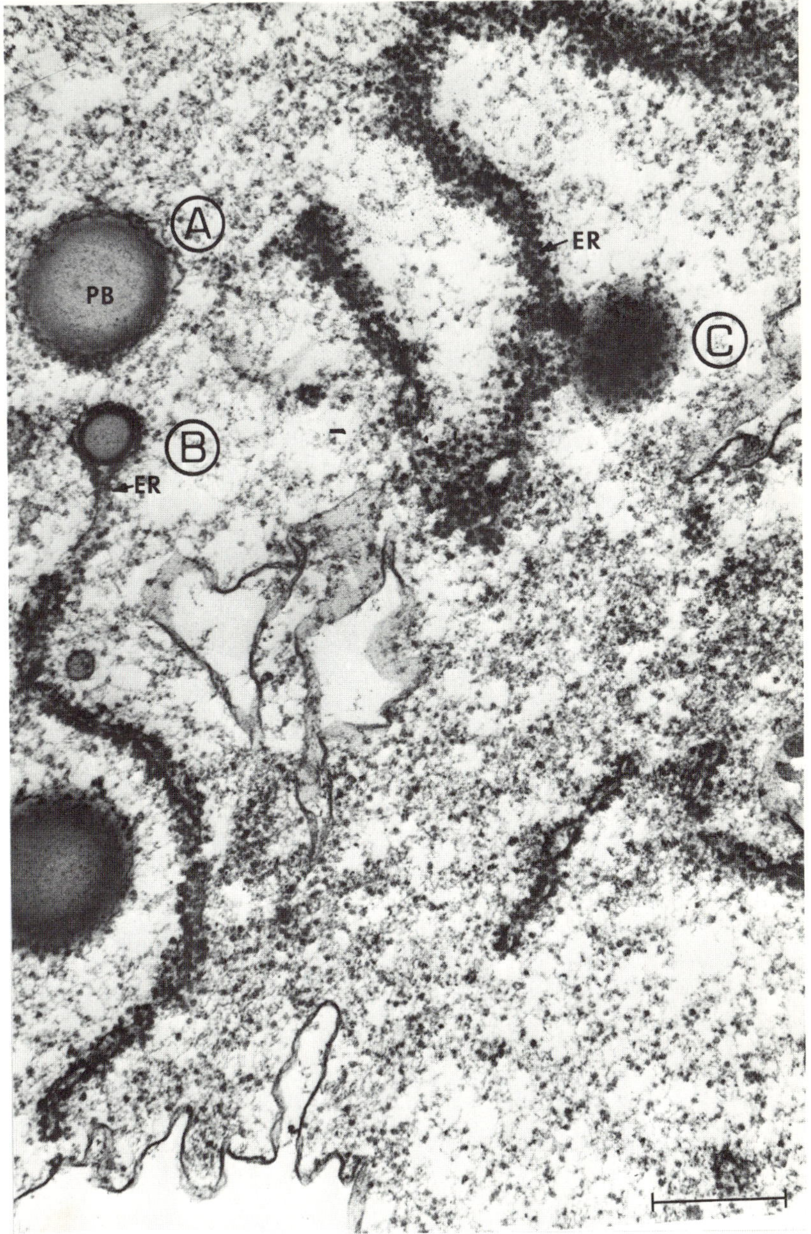

FIGURE 2. Electron micrograph of 19-day maize endosperm cell. PB, protein body; ER, rough endoplasmic reticulum; (A), (B), (C): see text. (Scale: bar = 0.5 μm).

membranes (Larkins and Dalby, 1975; Larkins *et al.*, 1976), this seemed like an important question to resolve.

To determine if membranes surrounding protein bodies had properties of RER, maize kernels were homogenized in a tris-MES buffer containing EDTA, and the 500 x g supernatant was analyzed on a continuous sucrose gradient (Larkins and Hurkman, 1978). The gradient was fractionated and each fraction was analyzed for cytochrome c reductase, which is the enzyme commonly used as a marker for endoplasmic reticulum (Leonard *et al.*, 1973).

There were three regions of cytochrome c reductase activity in the gradient (Fig. 3). The zone of highest activity had the greatest uv absorbance and corresponded to RER, as it was displaced to a greater density when magnesium was included in the gradient. The second zone of activity had a small uv absorbance and corresponded to mitochondrial membranes since it also had cytochrome c oxidase activity. The third peak of activity was associated with a light-scattering band of protein bodies. Although this experiment indicated that the membranes surrounding protein bodies have properties of ER, it does not prove that polysomes on protein body membranes are identical to those of the RER.

To answer this question, polysomes were isolated from RER and protein body membranes. Maize kernels were homogenized in polysome buffer and the 500 x g supernatant was analyzed on discontinuous sucrose gradients consisting of layers of 0.5, 1.5, and 2.0 M sucrose (Fig. 4). A RER-enriched fraction was isolated between the 0.5 and 1.5 M sucrose layers. It was determined that this fraction was free of protein bodies based upon examination of the preparation by electron microscopy and by the absence of zein polypeptides in alcohol extracts. A protein body-enriched fraction was isolated from between the 1.5 and 2.0 M sucrose layers. Polysomes prepared from these two fractions showed identical size distributions with polysomes containing 8 to 9 ribosomes per mRNA as the most abundant size classes. Of the total polysomes recovered, 60% were associated with protein bodies.

We tested for variability between these two polysome populations first by comparing the proportion of alcohol-soluble protein they synthesized. The purified polysomes were placed in a wheat germ cell-free protein synthesizing system (Marcu and Dudock, 1974) and the proportion of ^{14}C-leucine incorporated into 70% ethanol-soluble protein determined (Table I). There was slight variation in the total incorporation in each reaction, although they contained approximately equal amounts of polysomes. In the first experiment polysomes from the RER and polysomes from protein bodies were compared with polysomes from a mixed sample. In this experiment approximately 58% of the protein synthesized by RER polysomes was ethanol-soluble, 61% of that

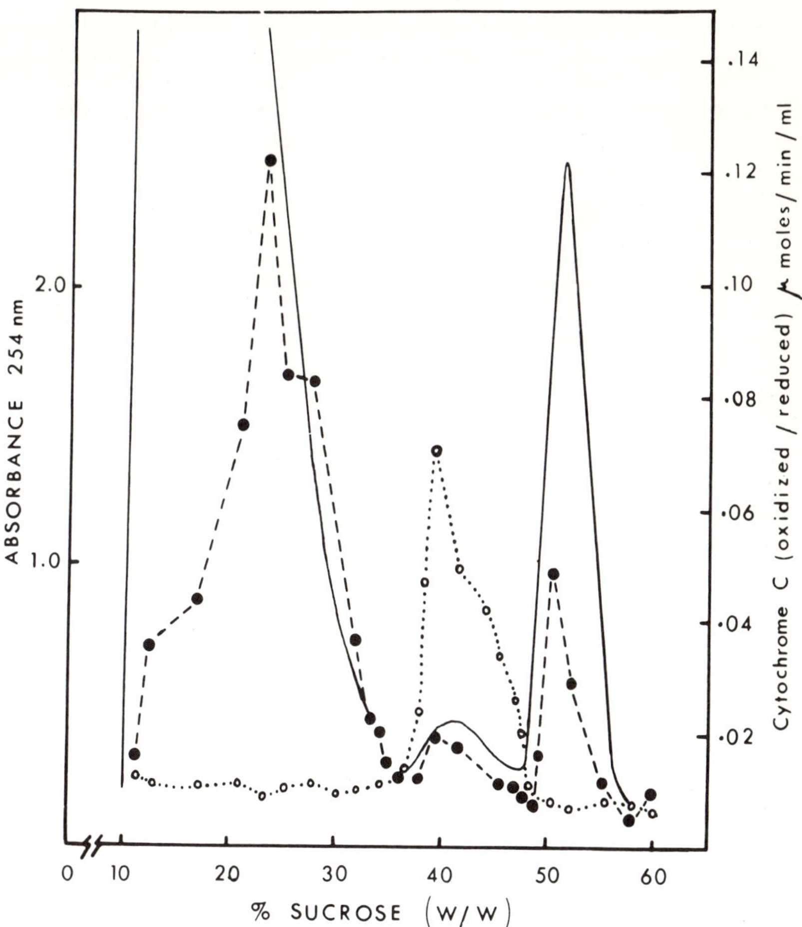

FIGURE 3. Distribution of cytochrome c reductase and cytochrome c oxidase activity after sucrose gradient centrifugation of maize endosperm extract. (————) absorbance at 260 nm; (--------) cytochrome c reductase activity; (........) cytochrome c oxidase activity (courtesy of The American Society of Plant Physiology).

synthesized by protein body polysomes was ethanol-soluble, and 70% of the protein synthesized by polysomes from a mixture of the two was ethanol-soluble. The difference between RER and protein body polysomes was not significant, and, surprisingly, polysomes from the mixed sample synthesized the greatest proportion of alcohol-soluble protein.

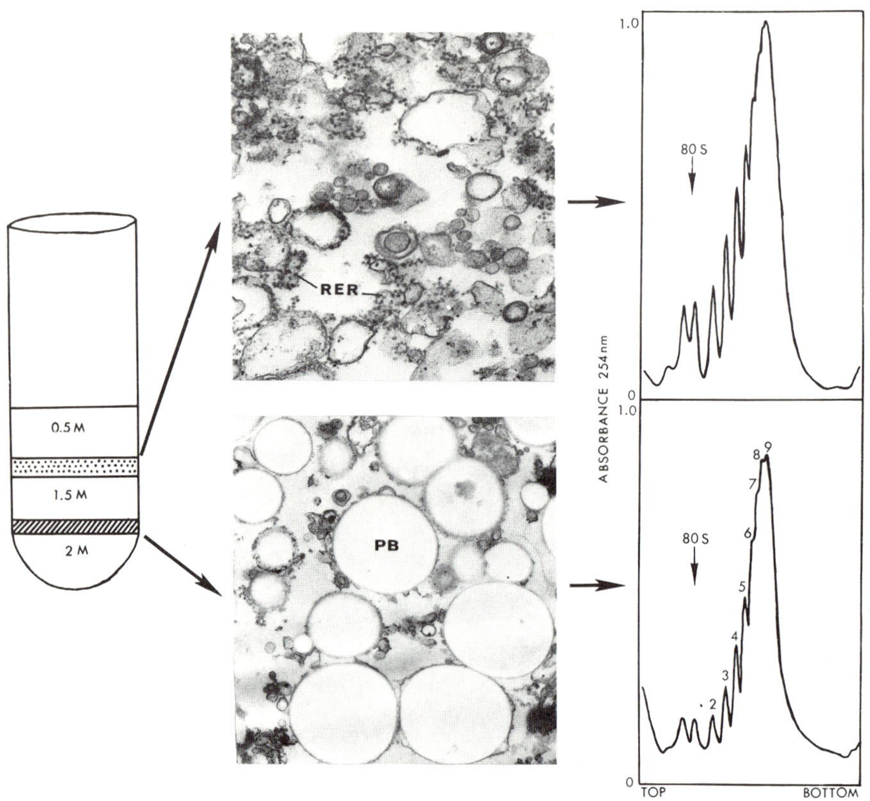

FIGURE 4. Separation of rough endoplasmic reticulum and protein bodies by discontinuous sucrose gradient centrifugation. Polysomes were released from RER and protein body membranes by treatment with 1% triton-X100. After pelleting through 2 M sucrose, polysomes were suspended in 100 μg/ml protease K and analyzed on 15-60% linear sucrose gradients.

To establish that this similarity was not a result of incomplete separation of RER and protein bodies with discontinuous sucrose gradients, we compared RER polysomes with polysomes from protein bodies isolated by continuous and discontinuous sucrose gradient centrifugation (Table I, Exp. 2). In this experiment the percentage of alcohol-soluble protein synthesized was as follows: RER polysomes, 45%; protein body polysomes from discontinuous sucrose gradients, 50%; and protein body polysomes from continuous sucrose gradients, 52%. Again the percentage of alcohol-soluble protein was similar in all polysome preparations.

TABLE I. Comparison of ^{14}C-Leucine Incorporation into Total and Ethanol-Soluble Protein Synthesized by Membrane-Bound Polyribosomes[a]

	Fraction A		Fraction B		Fractions A & B		Fraction PB	
	Total cpm	% EtOH-soluble	Total cpm	% EtOH-soluble	Total cpm	%EtOH-soluble	Total cpm	% EtOH-soluble
Exp. 1	116,000	58	106,000	61	92,000	70	---	---
Exp. 2	74,000	45	58,000	50	---	---	53,000	52

[a]Fractions A and B correspond to polyribosomes derived from rough endoplasmic reticulum and protein bodies, respectively, from discontinuous sucrose gradients. A mixture of A and B was obtained by isolating polyribosomes from membranes pelleted between 500 and 37,000 x g. Polyribosomes in fraction PB were obtained from protein bodies recovered from a continuous sucrose gradient. Each 50 µl assay contained approximately 0.85 A_{260} of polyribosomes in Experiment 1, and 0.62 A_{260} in Experiment 2. The radioactive counts were the average of triplicate assays and are expressed as cpm/50 µl reaction.

The analysis of alcohol-soluble proteins from these reactions by SDS-polyacrylamide gel electrophoresis showed that identical proteins were synthesized. But since this similarity may reflect selection for a certain class of proteins, we also analyzed the alcohol-insoluble proteins (Fig. 5). The labeling pattern of the 70% alcohol-insoluble proteins from the RER (Fig. 5, gel A), protein body polysomes from discontinuous sucrose gradients (Fig. 5, gel B), and protein body polysomes from continuous sucrose gradients (Fig. 5, gel PB) were virtually identical. Each contained a heterogeneous distribution of labeled proteins with prominent bands corresponding in molecular weight to the major zein proteins.

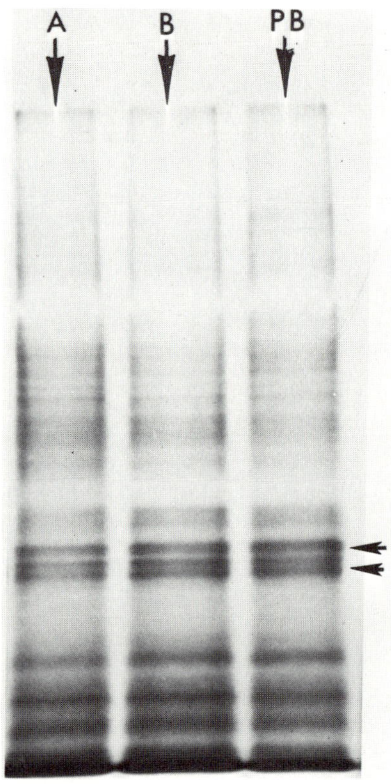

FIGURE 5. SDS-polyacrylamide gel analysis of alcohol-insoluble protein synthesized by membrane-bound polysomes from (A) RER, (B) protein bodies from discontinuous sucrose gradients, and (PB) protein bodies from continuous sucrose gradients. Arrows indicate the position of the major ethanol-soluble radioactive protein (courtesy of The American Society of Plant Physiology).

The similarity between protein body and RER membranes and the polysomes associated with them indicates that endosperm protein bodies form as nonspecific deposits within the RER. Homogenization of endosperm tissue results in fragmentation of RER, and because of the greater density of protein bodies, two fractions of membrane-bound polysomes are recovered. The 60% to 40% distribution of polysomes that we observed between RER and protein body membranes is therefore an artifact of tissue homogenization. A similar population of polysomes, hence mRNAs, exists in both fractions.

Figure 6 shows an analysis of the poly(A)-containing RNA isolated from the total membrane-bound polysome fraction. This sample was analyzed on a linear-log sucrose gradient in which the mRNA sedimented at approximately 13 S (Larkins *et al.*, 1976). The gradient was fractionated and after the mRNA was recovered, it was translated in the wheat germ cell-free protein synthesizing system. Figure 7 shows a fluorographic analysis of the products. The mRNA in fraction 10 directed the synthesis of neither of the major zein subunits. The mRNA

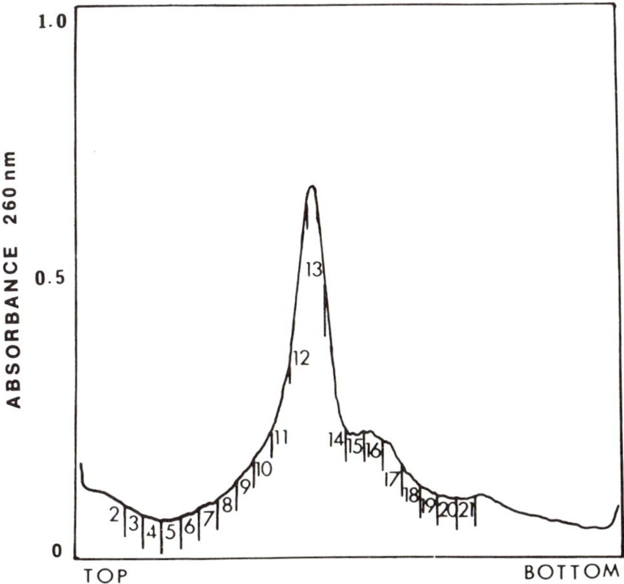

FIGURE 6. Analysis of poly(A)-containing RNA isolated from total membrane-bound polysomes on a linear-log sucrose gradient. The gradient was divided into 0.4 ml fractions and the RNA precipitated (courtesy of The American Society of Plant Physiology).

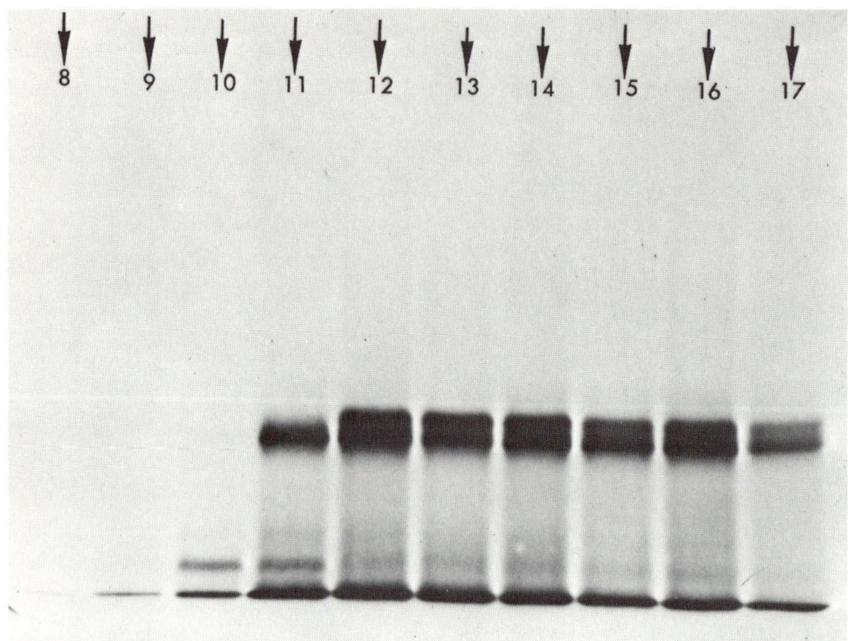

FIGURE 7. SDS-polyacrylamide gel analysis of the hot 5% trichloroacetic acid-insoluble proteins synthesized by fractionated mRNA. Sample numbers correspond to the gradient fractions designated in Fig. 6 (courtesy of The American Society of Plant Physiology).

in fraction 11 directed the synthesis of the smaller molecular weight zein subunit, while that in fraction 12 and several subsequent fractions directed the synthesis of both subunits. This result demonstrates that there are different but similar-sized mRNAs responsible for zein synthesis.

The results shown in Fig. 8 support the existence of separate but similar-sized mRNAs. In the opaque-2 mutant the 22,000 dalton zein component is suppressed (Lee *et al.*, 1976), but the mRNA from membrane-bound polysomes of opaque-2 has the same apparent molecular weight as that from normal maize (Jones, 1976). When the translation products of this mRNA (Fig. 8, gel D) were compared with those of fractionated normal mRNA (Fig. 8. gels A and B) or total normal mRNA (Fig. 8, gel C), it was apparent that mRNA from the opaque-2 mutant directed the synthesis of only the smaller molecular weight zein component.

Although it was previously reported that zein synthesized *in vitro* had a molecular weight similar to the native protein (Larkins and Dalby, 1975; Burr and Burr, 1976), from these

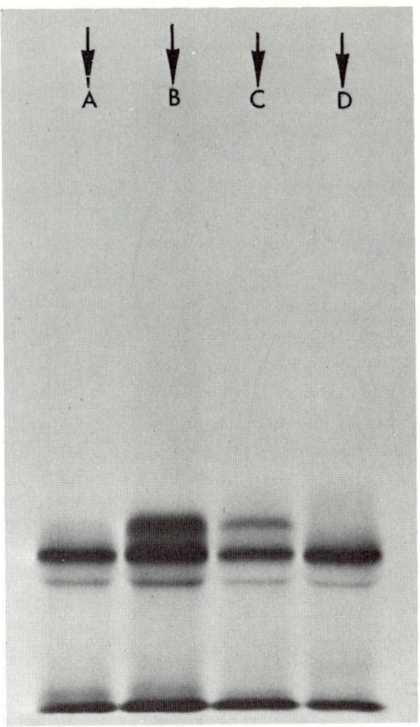

FIGURE 8. SDS-polyacrylamide gel analysis of the hot 5% trichloroacetic acid-insoluble proteins synthesized by (A,B) fractionated normal mRNA, (C) nonfractionated normal mRNA, and (D) nonfractionated opaque-2 mRNA.

fluorographic analyses we found that zein synthesized *in vitro* was slightly larger than the native protein. Compared to ØX 174 proteins (Fig. 9) the native proteins had molecular weights of 22,000 and 19,000, whereas the *in vitro* products had molecular weights of 24,000 and 21,000. This result suggested that the zein proteins are synthesized as precursors similar to proteins synthesized by membrane-bound polysomes in animal tissues (Shore and Tata, 1977). Since a portion of the N-terminal region of these proteins is removed during secretion through the endoplasmic reticulum in animal tissue (Jackson and Blobel, 1977), we compared the alcohol-soluble proteins synthesized by polysomes stripped from RER with those synthesized by intact RER. Gel A in Fig. 10 shows the products of purified membrane-bound polysomes, and gel B shows the products of intact RER added to the wheat germ cell-free protein synthesizing system. The proteins synthesized by intact RER were smaller in molecular weight than those synthesized by isolated polysomes.

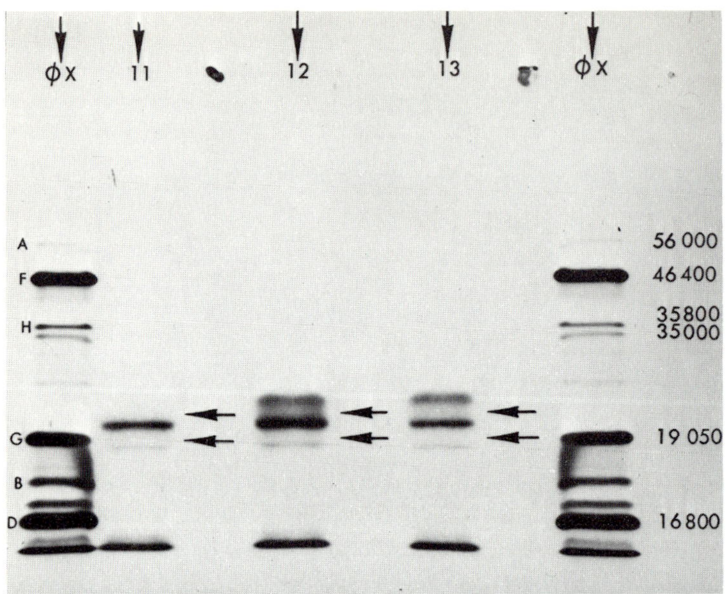

FIGURE 9. Comparative migration of proteins synthesized by fractionated mRNA and ØX 174 proteins. Horizontal arrows indicate the location of stained native zein proteins. The molecular weights of the ØX 174 proteins were based upon nucleotide sequence calculations of Sanger et al. (1977). (Courtesy of the American Society of Plant Physiology).

This is quite evident for the larger zein component. It is less evident for the smaller zein component, but this is primarily because of the more intense labeling of this protein by the purified polysome preparation. While we have insufficient evidence to conclude that a portion of the N-terminal region of the protein was removed, this alteration was very reproducible and did not appear to result from nonspecific protolytic activity.

Many eucaryotic mRNAs have a 7-methylguanosine-5'-phosphate (^{7}mGp) "cap" on the 5' end. The ^{7}mG and the penultimate nucleotide are joined through their 5'-hydroxyl groups to form a 5'-5' linkage which is inverted relative to the normal 3'-5' phosphodiester bonds (Shatkin, 1976). Although this structure has been found in a number of eucaryotic mRNAs, picornavirus and satellite tobacco necrosis virus (STNV) RNAs are not capped. Because translation of capped mRNAs in cell-free protein synthesizing systems is inhibited by the presence of

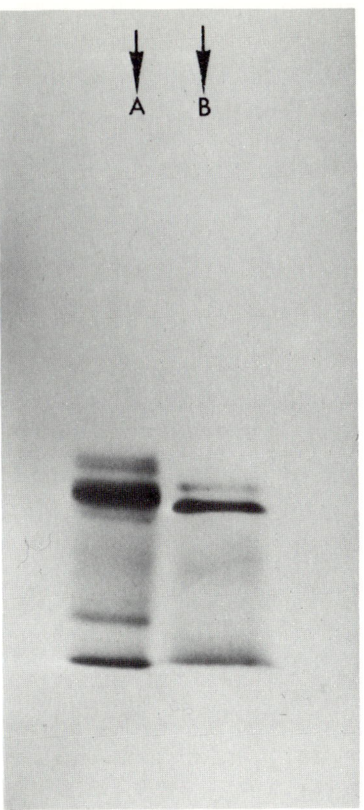

FIGURE 10. SDS-polyacrylamide gel analysis of alcohol-soluble proteins synthesized in vitro *(A) by membrane-bound polyribosomes isolated from RER, or (B) attached to RER (courtesy of The American Society of Plant Physiology).*

m^7Gp, the inhibited translation of mRNAs by m^7Gp has been cited as evidence for the presence of a capped structure (Sharma *et al.*, 1976).

Since zein mRNA is polyadenylated like other eucaryotic mRNAs (Larkins *et al.*, 1976) we attempted to determine if it also had a m^7G cap. Zein mRNA was translated in the wheat germ cell-free protein synthesizing system in the presence of varying concentrations of m^7Gp. Tobacco mosaic virus (TMV) RNA (a capped mRNA) and STNV RNA (a noncapped mRNA) were used as controls. Translation of TMV RNA was 98% inhibited by 0.2 mM ^7mGp, and this inhibition increased slightly at higher analog concentrations (Fig. 11). Translation of STNV RNA was 40% inhibited by 0.2 mM ^7mGp, and its translation was progressively inhibited at higher m^7Gp concentrations. The transla-

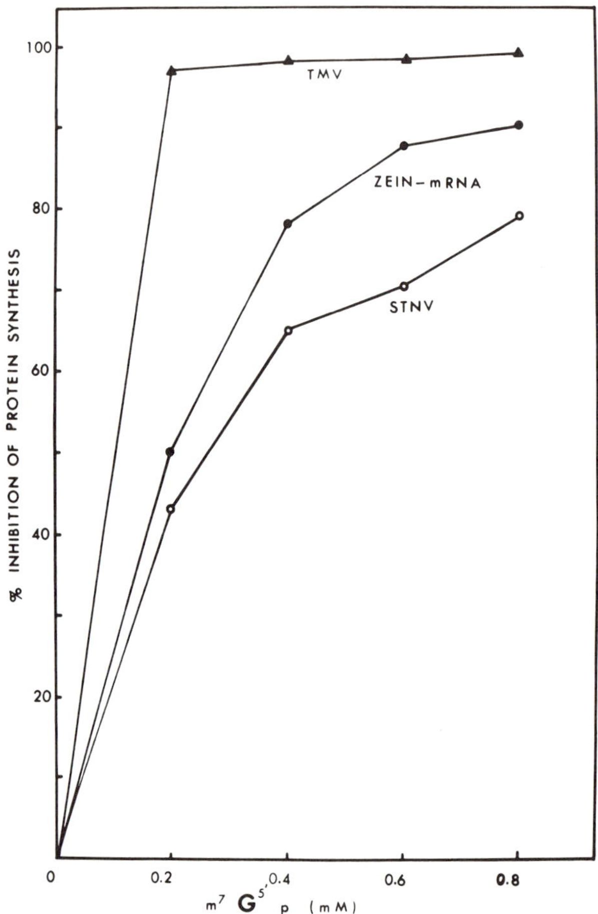

FIGURE 11. Inhibition of in vitro *translation of zein mRNA by 7-methylguanosine monophosphate. The wheat germ cell-free protein synthesizing system was prepared according to Marcu and Dudock (1974); the reactions contained 0.5 μg of TMV, 1.0 μg STNV, or 1.0 μg of zein mRNA.*

tion of zein mRNA was 50% inhibited by the lowest concentration of m^7Gp, which increased to 90% inhibition at higher m^7Gp concentrations. Although translation of zein mRNA was inhibited by m^7Gp, the degree of inhibition was intermediate between a capped (TMV) and a noncapped (STNV) mRNA.

It cannot be concluded from these results that the storage protein mRNA is capped. Although translation of the mRNA was inhibited by low concentrations of m^7Gp, the extent of inhibition was more similar to that of the uncapped mRNA.

Recently it has been reported that inhibited translation of capped mRNAs by m^7Gp is influenced by both the concentration and type of potassium salt in the cell-free translation system (Kemper and Stolarsky, 1977). Therefore, additional characterization is necessary to conclude safely that an mRNA is capped.

SUMMARY

Contrary to previous suggestions that protein bodies in maize endosperm are highly differentiated sites of storage protein synthesis, our results demonstrate that they form non-specifically as deposits within the RER. This conclusion is based on localization of cytochrome c reductase activity in protein body and RER membranes, similarity in the proportion of alcohol-soluble proteins synthesized by polysomes from RER and protein body membranes, and synthesis of identical alcohol-insoluble proteins by both fractions of polysomes. The mechanism of zein biosynthesis resembles the mechanism of secretory protein synthesis in some animal tissues. In both cases a "signal" sequence on the amino-terminus of the nascent polypeptide appears to initiate interaction between the ribosome and the RER resulting in the formation of a membrane-bound polyribosome. As the protein is discharged into the lumen of the RER, this signal sequence is removed.

We conclude that different but similar-sized mRNAs are responsible for synthesis of the two major zein components. This conclusion is based on partial separation of these mRNAs by sucrose gradient centrifugation, and synthesis of only one of the major zein components by mRNA from the opaque-2 mutant. Although these mRNAs have a poly(A)-rich region like many other eucaryotic mRNAs, there is at present insufficient data to conclude that the mRNA has a 7-methylguanosine cap at the 5' end of the mRNA.

ACKNOWLEDGMENTS

The authors wish to thank Dr. Abraham Marcus for the gift of TMV and STNV RNA, as well as for his useful criticisms and suggestions for the 7-methylguanosine-5'-phosphate experiments. Contribution from Purdue University Experiment Station, paper no. 7172.

REFERENCES

Burr, B., and Burr, F. A. (1976). Zein synthesis in maize endosperm by polyribosomes attached to protein bodies. *Proc. Nat. Acad. Sci. (U.S.) 74,* 515-519.

Burr, B., Burr, F. A., and Hannah, L. C. (1977). Localized synthesis of zein in maize endosperm. *Cereal Foods World 22,* 469.

Jackson, R. C., and Blobel, G. (1977). Post-translational cleavage of presecretory proteins with an extract of rough microsomes from dog pancreas containing signal peptidase activity. *Proc. Nat. Acad. Sci. (U.S.) 74,* 5598-5602.

Jones, R. A. (1976). Genetic regulation of storage protein biosynthesis in developing maize endosperm. Doctoral thesis. Purdue University, West Lafayette, Indiana.

Kemper, B., and Stolarsky, L. (1977). Dependence of potassium concentration on the inhibition of translation of messenger ribonucleic acid by 7-methylguanosine-5'-phosphate. *Biochem. 16,* 5676-5780.

Khoo, U., and Wolf, M. J. (1970). Origin and development of protein granules in maize endosperm. *Amer. J. Bot. 57,* 1042-1050.

Larkins, B. A., and Dalby, A. (1975). *In vitro* synthesis of zein-like protein by maize polyribosomes. *Biochem. Biophys. Res. Commun. 66,* 1048-1054.

Larkins, B. A., Jones, R. A., and Tsai, C. Y. (1976). Isolation and *in vitro* translation of zein messenger ribonucleic acid. *Biochem. 15,* 5506-5511.

Larkins, B. A., and Hurkman, W. J. (1978). Synthesis and deposition of zein in protein bodies of maize endosperm. *Plant Physiol. 62,* 256-263.

Lee, K. H., Jones, R. A., Dalby, A., and Tsai, C. Y. (1976). Genetic regulation of storage protein content in maize endosperm. *Biochem. Genet. 14,* 641-650.

Leonard, R. T., Hansen, D., and Hodges, T. K. (1973). Membrane-bound adenosine triphosphatase activities of oat roots. *Plant Physiol. 51,* 749-754.

Marcu, K., and Dudock, B. (1974). Characterization of a highly efficient protein synthesizing system derived from commercial wheat germ. *Nucleic Acids Res. 1,* 1385-1397.

Morton, R. K., and Raison, J. K. (1964). The separate incorporation of amino acids into storage and soluble proteins catalyzed by two independent systems isolated from developing wheat endosperm. *Biochem. J. 91,* 528-539.

Sanger, F., Air, G. M., Barrell, B. G., Brown, M. L., Caulson, A. R., Fiddes, J. C., Hutchison, C. A. III, Slocombe, P. M., and Smith, M. (1977). Nucleotide sequence of bacteriophage ØX 174 DNA. *Nature 265,* 687-695.

Sharma, O. K., Hruby, D. E., and Beezley, D. N. (1976). Inhibition of ovalbumin translation by 7-methylguanosine-5'-phosphate. *Biochem. Biophys. Res. Commun. 72,* 1392-1398.

Shatkin, A. J. (1976). Capping of eucaryotic mRNAs. *Cell 9,* 645-653.

Shore, G. C., and Tata, J. R. (1977). Functions for poly-ribosome-membrane interactions in protein synthesis. *Biochem. Biophys. Acta. 472,* 197-236.

THE PLANT SEED
Development, Preservation, and Germination

ISOLATION AND CHARACTERIZATION OF MESSENGER RNAs THAT CODE FOR THE SUBUNITS OF SOYBEAN SEED PROTEIN

*R. N. Beachy**
J. F. Thompson
J. T. Madison

U.S. Plant, Soil and Nutrition Laboratory
Ithaca, New York

Legume seeds are a major source of protein for many monogastric animals, and storage proteins comprise over half of the protein in these seeds. Since large quantities of storage proteins are synthesized in a specific organ, the seed, over a relatively short period of time, developing seeds have gained increasing interest as research tools for studies of gene regulation. The possibilities that such studies might lead to significant practical results cannot be overlooked.

As part of a program to understand the reasons for the marked differences in protein content between soybean (40-45% protein) and other legumes (20-25% protein), we measured the turnover rates of the storage proteins of soybean (*Glycine max* L., Merr) and garden peas (*Pisum sativum* L.). The very low rates of degradation of storage proteins (during seed development) of both pea and soybean demonstrated that the relatively low protein content of peas was not due to degradation (Madison, personal communication). The possibility that synthetic rates might account for protein levels led to an initial study in which DNA, RNA, and protein contents were measured in developing seeds of soybean, pea, peanut (*Arachis hypogaea* L.), and kidney bean (*Phaseolus vulgaris* L.) (Madison *et al.*, 1976). The lack of any correlation between total DNA and RNA, and protein levels led to an investigation of the messenger RNAs for storage proteins. Soybeans were chosen for the initial stages of this investigation because their storage proteins were reasonably well characterized.

**Present address: Department of Biology, Washington University, St. Louis, Missouri 63130*

ISBN 0-12-602050-7

THE PROTEINS INVOLVED

The review by Derbyshire *et al.* (1976) provides a good background on the storage proteins of soybean. Recently, workers in Japan have reported the amino acid analysis of the soybean storage proteins (Kitamura and Shibasaki, 1975; Thanh and Shibasaki, 1977) and data on dissociation and renaturation of the storage proteins (Kitamura *et al.*, 1977; Thanh, 1976).

When globulin proteins (those soluble in concentrated salt solutions) from soybean (cv. Provar) were subjected to sucrose density gradient sedimentation analysis according to the method of Hill and Breidenbach (1974), prominent peaks of protein of approximately 7S (conglycinin) and 11S (glycinin) were observed (Fig. 1). The 2S to 5S proteins (fractions 4 to 12 of Fig. 1) are believed to be involved in cell metabolism (mostly enzymes) and turn over at a relatively rapid rate, whereas the storage proteins (7S and 11S) are stable (Madison, personal communication). The 2S to 5S proteins are synthesized throughout all stages of seed development. When isolated cotyledons from seeds in early, mid, and late stages of development were cul-

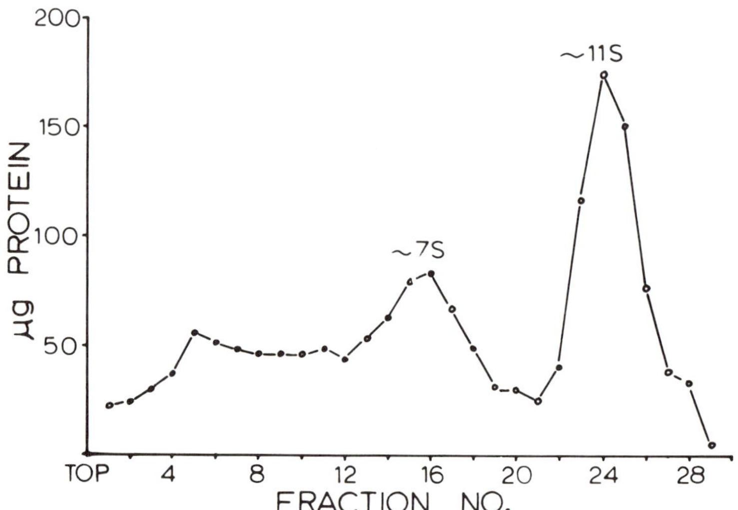

FIGURE 1. Sedimentation of soybean proteins in a sucrose density gradient. Proteins soluble in 0.4 M NaCl were extracted from soybeans and sedimented in 5 to 20% linear gradients as described by Hill and Breidenbach (1974). Protein content of individual fractions was determined by the Lowry method (Lowry et al., 1951). The approximate S values for the major storage proteins are given.

tured (Thompson *et al.*, 1977) in ^{3}H-leucine for 4 hr and the salt-soluble and acid-insoluble radioactivity was determined after sucrose gradient sedimentation, we found that the amounts of 7S and 11S proteins synthesized increased with the age of the cotyledon, while the level of incorporation into 2S to 5S protein was high throughout maturation (Fig. 2). The results demonstrated that isolation of mRNAs for the storage proteins would need to be carried out against the background of the mRNAs for these presumed nonstorage proteins.

Several groups of workers *(vide supra)* have extensively characterized the subunit composition of soybean storage proteins. After denaturation with sodium dodecyl sulfate (SDS)

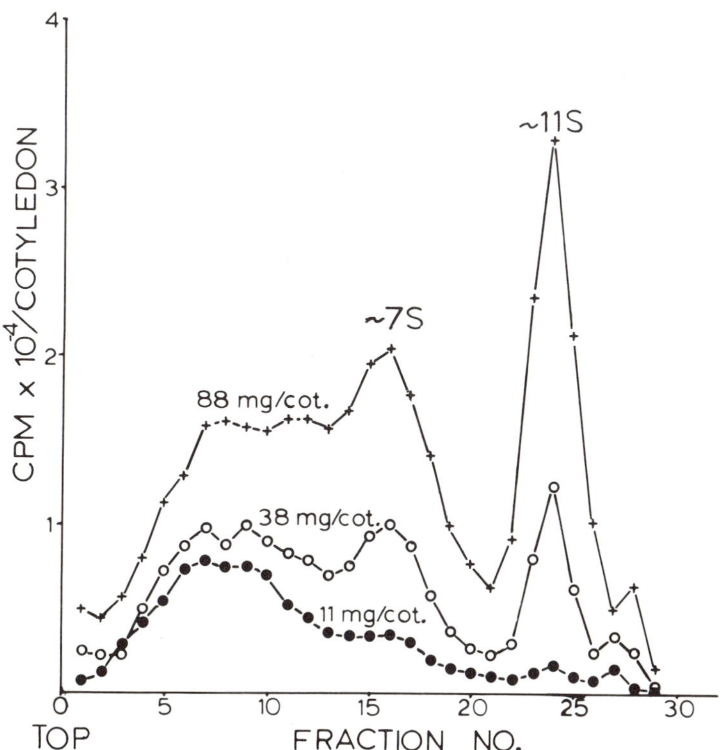

FIGURE 2. Sucrose gradient analysis of proteins synthesized in soybean cotyledons cultures. Seeds of three sizes were asceptically removed from their pods, embryos were excised, and single, weighed cotyledons were cultured in the presence of ^{3}H-leu for 4 hr (Thompson et al., 1977). Salt-soluble proteins were centrifuged on sucrose gradients and trichloroacetic acid insoluble radioactivity was determined. Fractions were taken from top to bottom.

and β-mercaptoethanol (MCE) the 11S protein breaks down into four basic subunits that, upon polyacrylamide gel electrophoresis (PAGE), have apparent molecular weights (M_r) of approximately 22,000 (Kitamura *et al.*, 1976; Draper and Catsmipoolas, 1977; Castro and Breidenbach, personal communication) and four acidic subunits with M_r 37,000 to 45,000 (Kitamura *et al.*, 1976; Castro and Breidenbach, personal communication). Our preparations of 11S protein exhibit similar subunit patterns (Fig. 3). Treatment of the 7S protein with SDS and MCE results in its breakdown into at least 3 subunits that Thanh and Shibasaki (1977) labeled α′, α, and β. Thanh supplied us with samples of purified subunits that coelectrophoresed with the subunits of 7S protein which we isolated (Fig. 3). In an effort to determine if one particular state of seed development might be advantageous over another as a source of a particular messenger RNA (mRNA) we carried out experiments to determine the stage of development at which these subunits were produced. Proteins extracted from seeds of various weights, ranging from immature (29 mg, fresh weight) through mature (dried), were subjected to sucrose density gradient centrifugation, and

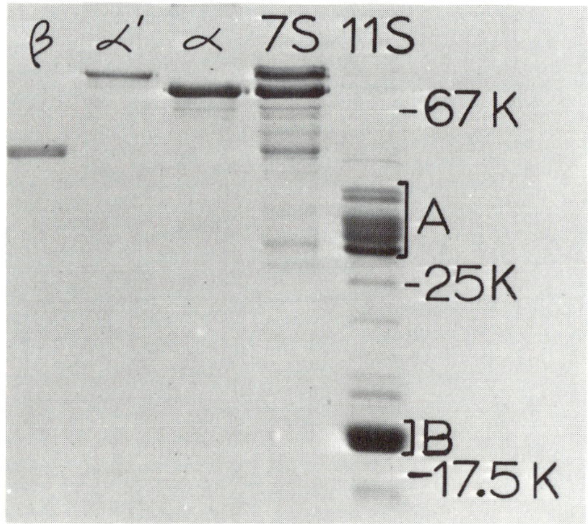

FIGURE 3. Subunit analysis of soybean 7S and 11S proteins. 7S and 11S proteins were disrupted with SDS and MCE and subjected to electrophoresis on a discontinuous 5 and 12% polyacrylamide slab gel according to Maizel (1971). Samples of purified 7S subunits, α′, α, and β were provided by Thanh (Tohoku University, Japan). The 11S protein subunits group as acidic (A) and basic (B) subunits. On the far right the positions to which three molecular weight markers migrated are indicated. The proteins were stained with Coomassie Brilliant Blue.

single peak fractions of the 7S or 11S proteins (10 to 15 μg per sample) were disrupted with SDS and MCE and subjected to electrophoresis and staining (Fig. 4). During the early stages of seed development, 7S proteins contain a predominating amount of α subunit and unidentified bands with electrophoretic migrations intermediate between α and β subunits; the α′ and β subunits appear later. The acidic and basic subunits of the 11S protein appear more or less simultaneously when the seed reaches about 71 mg fresh weight. The results indicated that the mRNAs for the subunits of the storage proteins could be isolated from seeds in the mid to late stages of development.

ISOLATION AND TRANSLATION OF THE MESSENGERS

All messenger RNA preparations were begun by purifying polyribosomes from harvested seeds that were frozen in liquid nitrogen and stored at -20° or -85°C. The polyribosome extraction buffer was the high pH-high KCl buffer previously described (Beachy *et al.*, 1978). Polyribosomes from seeds in early

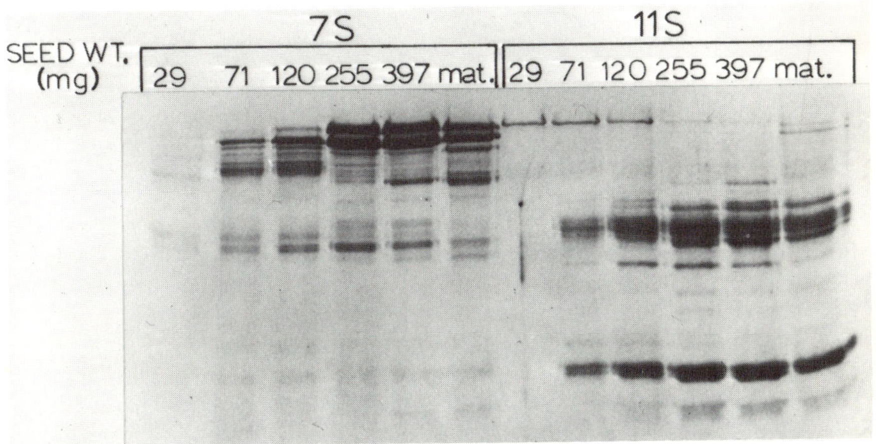

FIGURE 4. Subunit composition of the 7S and 11S proteins at different stages of seed development. Seeds of varying developmental stages from immature (29 mg fresh weight) to nearly mature, but not yet desiccated (397 mg), and mature (mat.) were extracted in high salt-containing buffer and subjected to sucrose density gradient centrifugation. Single peak fractions of the 7S or 11S proteins were treated with SDS and MCE and electrophoresed as described in Fig. 3.

(<150 mg/seed), mid (150 to 250 mg/seed), and late (>250 mg/seed) stages of development gave similar sedimentation patterns when centrifuged in 12.5 to 50% sucrose density gradients (Fig. 5). For mechanical reasons seeds that weighed 150 to 250 mg were routinely used for polyribosome isolation. For partial purification of mRNA, polyribosomal RNA was subjected to affinity chromatography on oligo dT-cellulose after SDS-disruption of the polyribosomes (Krystosek *et al.*, 1975), or after extraction with SDS (at 65°C) and phenol. Both intact polyribosomes and poly(A)-containing mRNA (that which bound to oligo dT-cellulose in the presence of 0.5 M NaCl) were used in *in vitro* protein synthesizing reactions containing wheat germ 23,000 x g supernatant (S-23) prepared according to Bruening *et al.* (1976). The optimal K^+ and Mg^{++} concentrations for *in vitro* translation differed for the two sources of mRNA (Table I). Experiments with the protein synthesis initiation inhibitor aurintricarboxylic acid (Marcus *et al.*, 1970) demonstrated

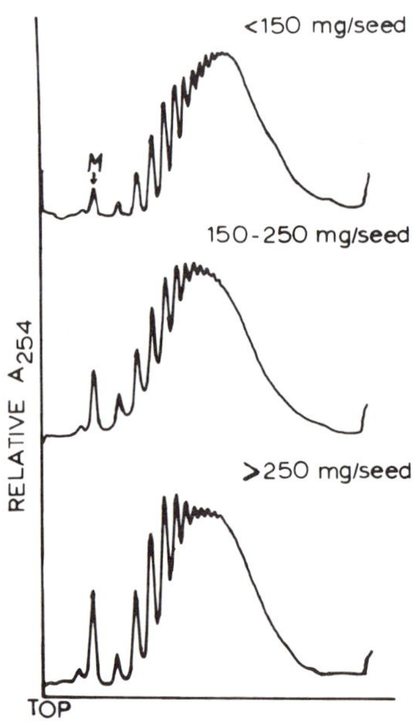

FIGURE 5. Sucrose density gradient analysis of polyribosomes extracted from soybean seeds. Soybeans in early (<150 mg/seed), mid (150-250 mg/seed), and late (>250 mg/seed) stages of development were extracted in high KCl-high pH buffer (Beachy et al., 1978) and centrifuged in 12.5 to 50% sucrose density gradients. Gradients were monitored at 254 nm with an ISCO gradient monitor. M = monosome.

*TABLE I. Optimal Mg++ and K+ Concentrations for the In Vitro Translation of Soybean Polyribosomes and Poly(A)+ RNA**

	Polyribosomes	Poly(A)+ RNA
Mg^{++}, + 0.4 mM spermidine	2.5 mM	1.95 mM
Mg^{++}, w/o spermidine	4.0 mM	Not done
K^{+}	60 to 90 mM	105 to 140 mM

**Wheat germ extract (23,000 x g supernatant) was prepared from commercial wheat germ as described by Bruening et al. (1976). Optimal concentrations for Mg^{++} (added as magnesium acetate) in the presence or absence of 0.4 mM spermidine and K^{+} (added as potassium acetate) were determined by assaying the incorporation of ^{3}H-leucine into hot TCA-insoluble radioactivity in reaction mixtures containing varying amounts of these salts (Beachy et al., 1978). Messenger RNA was added either as total polyribosomes or as poly(A)+ RNA (mRNA bound to oligo dT-cellulose in the presence of 0.5 M NaCl).*

that when polyribosomes were added, ^{3}H-leucine was incorporated only into nascent polypeptide chains isolated with the polyribosomes (Beachy *et al.*, 1978, Fig. 2C), i.e., no new initiation had occurred. Incorporation of ^{3}H-leucine continued for about 20 min when polyribosomes were added to reaction mixtures, but for more than 90 min when poly(A)-containing RNA (poly(A)+ RNA) was added. In experiments with the poly(A)+ RNA the translation reactions were stimulated in a linear fashion by the addition of from 0.2 to 2.0 μg of RNA per 50 μl reaction mixture (data not given).

The *in vitro* translation of poly(A)+ RNA was inhibited 83% by the addition of 0.25 mM 7-methylguanosine 5'-monophosphate to reaction mixtures, and by 95% when 1.0 mM was added; there was no effect when 7-methylguanosine was added (Beachy *et al.*, 1978, Fig. 7). The results suggest that, like many eucaryotic mRNAs, soybean mRNAs are "capped" at their 5' ends (Lodish and Rose, 1977).

The products of the translation reactions were subjected to SDS and MCE treatment followed by PAGE on the same slab gel with subunits of 7S and 11S proteins isolated from cotyledons that were radiolabeled in culture. Radioactive bands were detected by fluorography. The products of the *in vitro* translation reactions were complex, with numerous bands that did not correspond in electrophoretic migration to the subunits of either the 7S or the 11S proteins. Because developing cotyledons synthesize large amounts of nonstorage proteins (see Fig. 2), such a finding is not unexpected. Other more prominent bands had electrophoretic mobilities identical to the standards, suggesting that some or all of the subunits of the storage proteins had been synthesized *in vitro*. There were no qualitative differences between the products synthesized in response to adding polyribosomes or poly(A)+ RNA (Beachy *et al.*, 1978, Fig. 5) indicating that the purification procedure had not to that point caused a loss of specific mRNA activity. As expected, the high molecular weight protein subunits were synthesized on heavy polyribosomes (Beachy *et al.*, 1978, Fig. 5).

The products from the translation of mRNA that did not bind to oligo dT-cellulose (probably due to shortened segments of polyadenylate, thereby preventing tight binding to oligo dT) did not differ from the products of the poly(A)+ RNA (data not shown).

It has been well documented (Larkins and Dalby, 1976; Burr and Burr, 1976) that polyribosomes containing zein mRNA are associated with membrane-bound protein bodies of corn. Several experiments were done to determine if "membrane associated" polyribosomes (representing, in these experiments, less than 35% of the polyribosomes extracted) of soybean synthesized different polypeptides than did the "free" polyribosomes when both were translated *in vitro*. No differences were found in these experiments. However, Barton and Cherry (personal communication) reported that several high M_r subunits (unidentified, but possibly subunits of the 7S protein) were translated *in vitro* when membrane associated polyribosomes were added as mRNA but not when free polyribosomes were added. In their experiments approximately 50% of the polyribosomes were associated with a membrane fraction. Whether these apparently conflicting results represent real or procedural differences is at present unclear. Since the extraction conditions used in our experiments were high in salts and thus possibly deleterious to the integrity of membrane associated RNAs, to refer to polyribosomes as "free" or "membrane associated" may be descriptions that are more operational than functional. In the addendum to this article by Burr and Beachy are shown electron micrographs of protein bodies from soybean cotyledons.

FURTHER CHARACTERIZATION OF THE RNAs

The poly(A)+ RNA has been subjected to sucrose density gradient analysis, most successfully on linear 5 to 20% sucrose in 99% dimethyl sulfoxide (DMSO) (Bishop *et al.*, 1969). Figure 6 presents data from three preparations of poly(A)+ RNA (in these experiments large amounts of rRNA contaminated the mRNA preparations) including the O.D. scan at 280 nm (upper section) and the messenger activity of the ethanol-precipitable RNA in

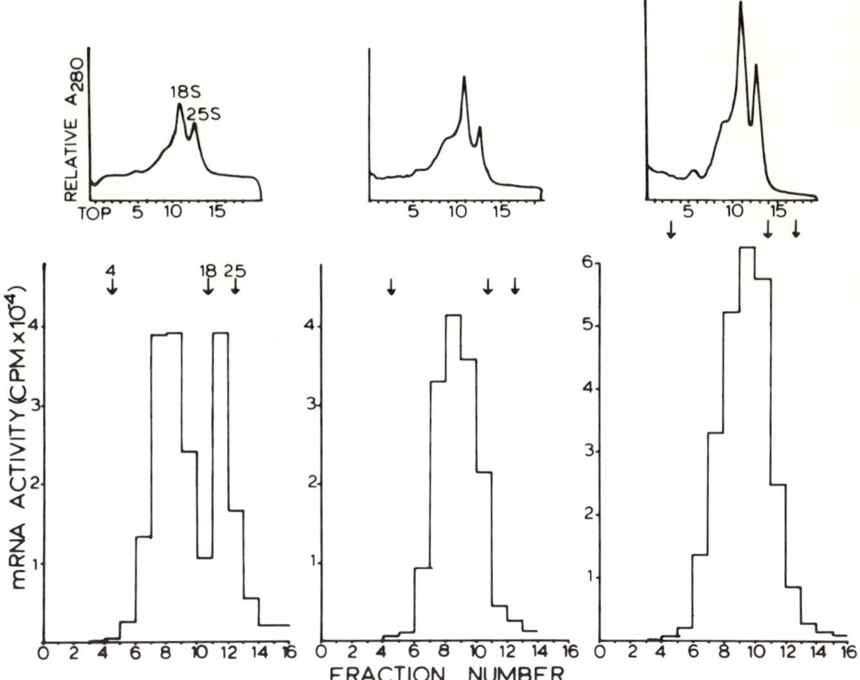

FIGURE 6. Sedimentation of poly(A)+ RNA in sucrose density gradients. RNAs which bound to oligo dT-cellulose in 0.5 M NaCl and subsequently eluted in a no-salt buffer were centrifuged in 5 to 20% sucrose density gradients prepared in 99% DMSO (Bishop et al., 1969), and monitored at 280 nm with an ISCO gradient monitor (upper portions). Messenger RNA activity was determined by in vitro *translation with the wheat germ system. The results from three different preparations of mRNA are given. Arrows indicate the positions to which 4S, 18S, and 25S RNAs sedimented in the gradients.*

each fraction (lower section). Messenger RNA activity ranged from less than 10S to about 24S and varied between experiments, probably as a result of inconsistent seed growing or seed processing procedures. For example, the data in Fig. 6, left panel, indicate that a large amount of mRNA activity sedimented between 18 and 25S, whereas the center and right panels had considerably less activity in that region. The products resulting from the *in vitro* translation in another experiment are given in Fig. 7 (note that more fractions were taken in this experiment than in those shown in Fig. 6). Subunits of the 7S and 11S proteins are also given. Apparently the mRNAs coding

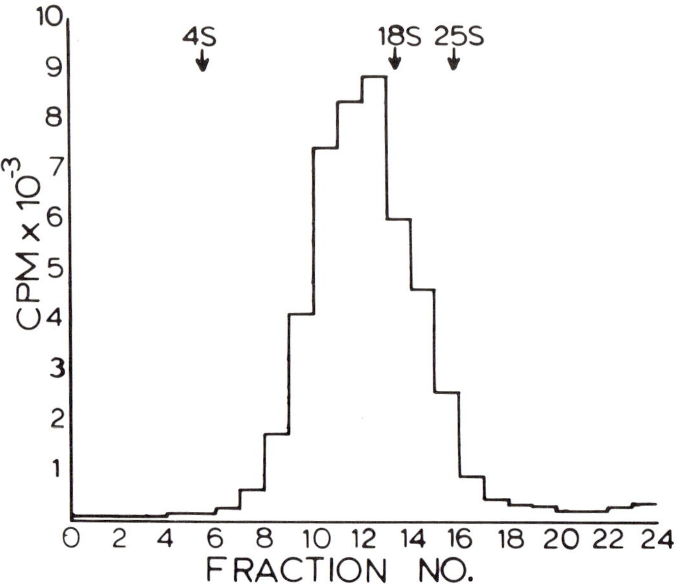

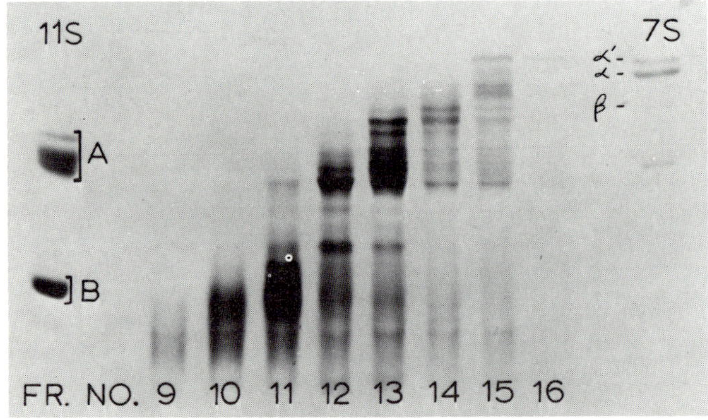

FIGURE 7. Analysis of the products resulting from in vitro *translation of poly(A)+ RNA. RNA fractionated on DMSO containing SDG (as described in Fig. 6) were translated* in vitro *(upper portion); aliquots of the reaction mixtures were treated with SDS and MCE, electrophoresed on a slab gel as described in Fig. 3, and radioactivity was detected by fluorography (lower portion). 11S and 7S proteins labeled in cotyledon culture were electrophoresed on the same slab gel. A = acidic subunits, B = basic subunits; α′, α, and β are designated subunits of 7S protein. Fraction numbers in the lower portion of the figure correspond to those in the upper portion.*

for the 11S basic subunits are found primarily in fractions 10 and 11, RNAs for the 11S acidic subunits in fractions 12 and 13, RNA for the 7S α subunit in fractions 14 and 15, and RNA for the 7S α′ subunit in fractions 15 and 16. In an early attempt to characterize the RNAs further, we subjected aliquots of the RNAs found in fractions of a DMSO-sucrose density gradient as shown in Fig. 6 to electrophoresis in polyacrylamide gels containing 2.0% polyacrylamide plus 0.5% agarose with 0.2% SDS (Loening, 1967). Gels were scanned at 260 nm, and the resulting tracings are given in Fig. 8. The M_r of the RNA increases with higher fraction numbers, as expected, and each fraction contains a heterodisperse population of molecules. Only in

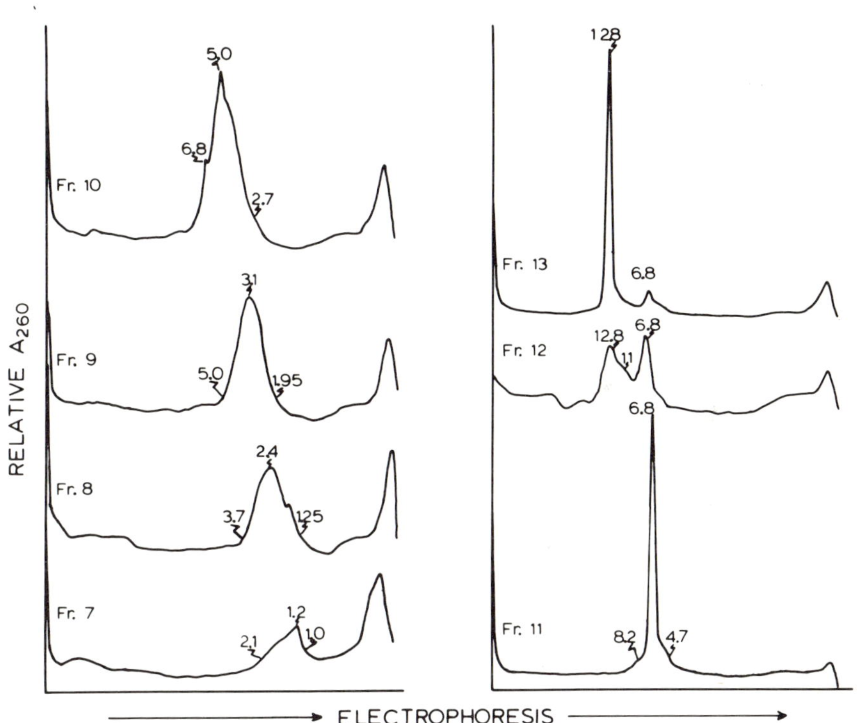

FIGURE 8. Gel electrophoretic analysis of the poly(A)+ RNAs. After DMSO-sucrose density gradient centrifugation, aliquots of the RNAs in fractions 7 through 13 were electrophoresed in 2.0% polyacrylamide containing 0.5% agarose and 0.2% SDS (Loening, 1967). Gels were scanned at 260 nm. The numbers given are molecular weights (x10^5) determined from a standard curve constructed with 25S and 18S soybean ribosomal RNAs, 23S and 18S soybean ribosomal RNAs, 23S and 16S E. coli *ribosomal RNAs and yeast tRNAs.*

fractions 11 to 13, containing ribosomal RNAs, are prominent peaks of RNA evident. Thus, even though the predominant polypeptides synthesized *in vitro* in response to the addition of RNA of fractions 8 and 9 comigrate with the basic subunits of 11S protein (data not shown), no corresponding predominant species of RNA is evident. Additional experiments with formamide or methyl mercury hydroxide as denaturants are currently underway to attempt further purification and characterization of the mRNAs involved. Preliminary results indicate that the latter technique, followed by electroelution of the RNA, will be useful in purifying specific RNAs.

VERIFICATION OF THE *IN VITRO* PRODUCT

We carried out several types of experiments to verify that the products synthesized *in vitro* are similar or identical to the standard proteins. In repeated experiments precise coincidence of migration of the *in vivo* and the *in vitro* polypeptides have been achieved for the 11S basic and acidic subunits, and for the α' and α subunits of the 7S protein. Since such analysis is not conclusive, however, authentic 11S basic subunits (labeled with ^{14}C-leucine) and polypeptides synthesized *in vitro* (labeled with ^{3}H-leucine) that have electrophoretic mobilities identical to those of the basic peptides have been electrophoretically eluted from gel pieces. We combined the eluted ^{14}C- and ^{3}H-labeled polypeptides, then digested them with trypsin and subjected them to ion-exchange chromatography (Vogt *et al.*, 1975) and determined the distribution of radioactivity. The data varied between experiments but most of the peaks of ^{14}C and ^{3}H were coincident (Beachy and Barton, unpublished). Some peaks, however, did not coincide, and we are at present unsure of the reasons. The data are consistent, however, and indicate that the RNA for at least some of the storage protein subunits has been isolated and translated *in vitro*.

NOW WHAT?

A great deal of interest has developed in the use of seed proteins as model systems in molecular biology. Other papers in this volume show that the work on *Phaseolus vulgaris* and *Zea mays* has progressed very well in the last few years. Other laboratories have begun the use of *Pisum sativum* (Higgins and Spencer, 1977), *Avena sativa* (Luthe and Peterson, 1977), and *Hordeum vulgaris* (Brandt and Ingverson, 1976; Miflin *et al.*, 1977). In each of these, polyribosomes and/or messenger RNAs have been translated *in vitro* to produce the expected storage protein or subunits thereof.

Availability of purified mRNA permits a multitude of experimental designs, some of which are presently under way, others of which are in the planning stages. Recent technological advances (such as those discussed in Scott and Werner, 1977) in nucleic acid research will permit the production of specific molecular probes that will be invaluable in studies of development. Other studies will explore factors, such as environment, nutrition, and pathology, that affect gene expression and possible ways for control of these factors. Such studies will also be facilitated by these recently developed technologies. In theory, the potentials for exploring basic aspects of plant biology are enormous. In practice, "we" are probably the limiting factor.

SUMMARY

The storage proteins of soybean characteristically sediment in sucrose density gradients with coefficients of approximately 7S (conglycinin) and 11S (glycinin). With sodium dodecyl sulfate and β-mercaptoethanol treatment the 7S protein breaks down into at least three polypeptide subunits, while the 11S protein breaks into four basic and four acidic subunits. Messenger RNAs have been isolated from polyribosomes taken from developing soybean seeds and after affinity chromatography on oligo dT-cellulose and sucrose density gradient centrifugation (under denaturing conditions) were translated in a wheat germ cell-free protein synthesizing system. Messenger RNA activities ranged from approximately 10S to 24S. Some, but not all, of the polypeptides synthesized *in vitro* had electrophoretic mobilities identical with some of the subunits of the storage proteins. Preliminary evidence from tryptic fragment analysis suggested that the basic subunits of the 11S protein are synthesized *in vitro*; other products have not béen comparably characterized. Polyacrylamide gel electrophoretic analysis of the poly(A)+ RNAs showed a widely heterodisperse population of mRNAs with no predominating species.

ACKNOWLEDGMENTS

We wish to thank Drs. Francis and Ben Burr for their encouragement and helpful discussions during the course of this research.

REFERENCES

Beachy, R. N., Thompson, J. F., and Madison, J. T. (1978). Isolation of polyribosomes and messenger RNA active in *in vitro* synthesis of soybean seed proteins. *Plant Physiol. 61,* 139-144.

Bishop, J. M., Koch, G., Evans, B., and Merriman, M. (1969). Polio-virus replicative intermediate: Structural basis of infectivity. *J. Mol. Biol. 46,* 235-249.

Brandt, A., and Ingverson, J. (1976). *In vitro* synthesis of barley endosperm proteins on wild type and mutant templates. *Carlsberg Res. Commun. 41,* 311-320.

Bruening, G., Beachy, R. N., Scalla, R., and Zaitlin, M. (1976). *In vitro* and *in vivo* translation of the ribonucleic acids of a cowpea strain of tobacco mosaic virus. *Virology 71,* 498-517.

Burr, B., and Burr, F. (1976). Zein synthesis in maize endosperm by polyribosomes attached to protein bodies. *Proc. Nat. Acad. Sci. U.S. 73,* 515-519.

Derbyshire, E., Wright, D. J., and Boulter, D. (1976). Legumin and vicilin, storage proteins of legume seeds. *Phytochemistry 15,* 3-24.

Draper, M., and Catsimpoolas, N. (1977). Isolation of the acidic and basic subunits of glycinin. *Phytochemistry 16,* 25-27.

Higgins, T. J. V., and Spencer, D. (1977). Cell-free synthesis of pea seed proteins. *Plant Physiol. 60,* 655-661.

Hill, J. E., and Breidenbach, R. W. (1974). Proteins of soybean seeds I. Isolation and characterization of the major components. *Plant Physiol. 53,* 742-746.

Kitamura, K., and Shibasaki, K. (1975). Isolation and some physicochemical properties of the acidic subunits of soybean 11S globulin. *Agr. Biol. Chem. (Tokyo) 39,* 945-951.

Kitamura, K., Takagi, T., and Shibasaki, K. (1976). Subunit structure of soybean 11S globulin. *Agr. Biol. Chem. (Tokyo) 40,* 1837-1844.

Kitamura, K., Takagi, T., and Shibasaki, K. (1977). Renaturation of soybean 11S globulin. *Agr. Biol. Chem. (Tokyo) 41,* 833-840.

Krystosek, A., Cawthon, M. L., and Kabat, D. (1975). Improved methods for purification and assay of eukaryotic messenger ribonucleic acids and ribosomes. Quantitative analysis of their interaction in a fractionated reticulocyte cell-free system. *J. Biol. Chem. 250,* 6077-6084.

Larkins, B. A., and Dalby, A. (1976). *In vitro* synthesis of zein-like protein by maize polyribosomes. *Biochem. Biophys. Res. Commun. 66,* 1048-1054.

Lodish, H. F., and Rose, J. K. (1977). Relative importance of 7-methylguanosine in ribosome binding and translation of vesicular stomatitis virus mRNA in wheat germ and reticulocyte cell-free systems. *J. Biol. Chem. 252,* 1181-1188.

Loening, U. E. (1967). The fractionation of high-molecular-weight ribonucleic acid by polyacrylamide-gel electrophoresis. *Biochem. J. 102,* 251-257.

Lowry, O. H., Rosebrough, N. J., Farr, A. L., and Randall, R. J. (1951). Protein measurement with the Folin phenol reagent. *J. Biol. Chem. 193,* 265-275.

Luthe, D. W., and Peterson, D. M. (1977). Cell-free synthesis of globulin in developing oat (*Avena sativa* L.) seeds. *Plant Physiol. 59,* 836-841.

Madison, J. T., Thompson, J. F., and Muenster, A. M. (1976). Deoxyribonucleic acid, ribonucleic acid, protein and uncombined amino acid content of legume seeds during embryogeny. *Ann. Bot. 40,* 745-756.

Maizel, J. V. (1971). Polyacrylamide gel electrophoresis of viral proteins. *In* "Methods in Virology," Vol. 1 (K. Maromarosch and H. Koprowski, eds.), pp. 179-246. Academic Press, New York.

Marcus, A., Bewley, J. D., and Weeks, D. P. (1970). Aurintricarboxylic acid and initiation factors of wheat embryo. *Science 167,* 1735-1736.

Miflin, B. J., Fox., J. E., and Shewry, P. R. (1977). The *in vitro* synthesis of barley storage proteins. *Plant Physiol. 59 (Suppl),* 109 (Abstr).

Scott, W. A., and Werner, R., eds. (1977). "Molecular Cloning of Recombinant DNA: Miami Winter Symposia," Vol. 13. Academic Press, New York.

Thanh, V. H. (1976). Beta-conglycinin from soybean proteins. The molecular properties and subunit structure. Ph.D. Thesis, Tohoku University, Sendai, Japan.

Thanh, V. H., and Shibasaki, K. (1977). Beta-conglycinin from soybean proteins. Isolation and immunological and physico-chemical properties of the monomeric forms. *Biochim. Biophys. Acta 490,* 370-384.

Thompson, J. F., Madison, J. T., and Muenster, A. M. (1977). *In vitro* culture of immature cotyledons of soya bean *(Glycine max* L. Merr). *Ann. Bot. 41,* 29-39.

Vogt, V. M., Eisenman, R., and Diggelmann, H. (1975). Generation of avian myeloblastosis virus structural proteins by protolytic cleavage of a precursor polypeptide. *J. Mol. Biol. 96,* 471-493.

Addendum: Protein Bodies of the Soybean Cotyledon

Frances A. Burr

Biology Department
Brookhaven National Laboratory
Upton, New York

R. N. Beachy

U.S. Plant, Soil and Nutrition Laboratory
Ithaca, New York
Present address: Department of Biology
Washington University
St. Louis, Missouri

In Figs. 1 and 2 are shown the protein bodies of the soybean cotyledon at two stages of cotyledon development.

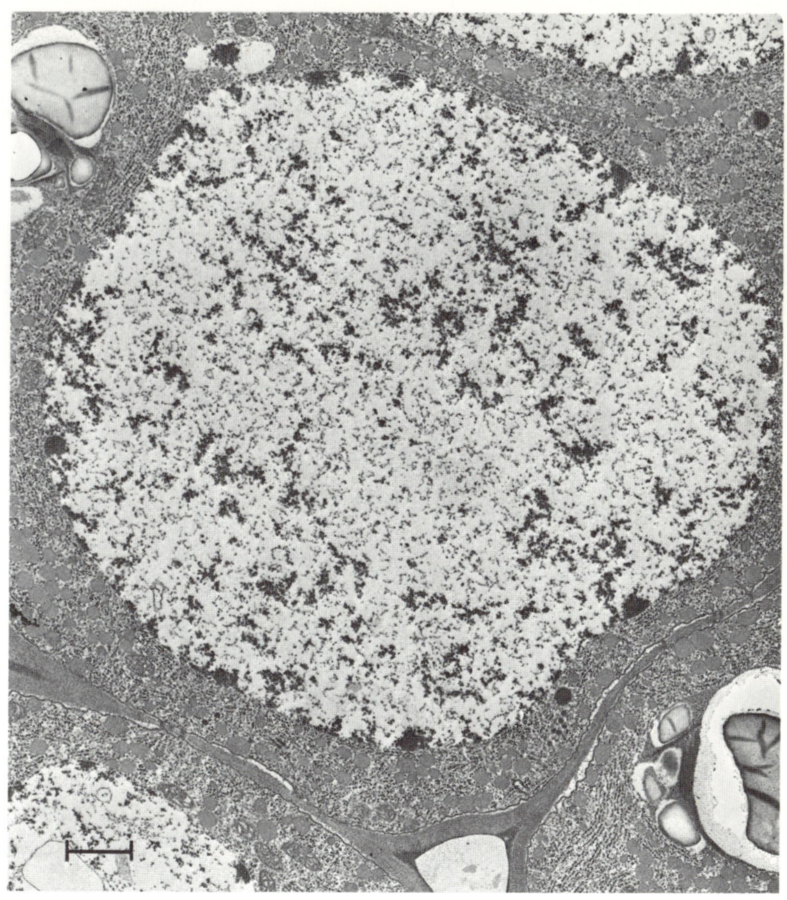

FIGURE 1. A developing protein body from 1 cm long cotyledons of soybean, cultivar Lee. The protein body is limited by a single membrane (with no attached polysomes). The flocculent material in the lumen represents the beginnings of storage protein accumulation. It is believed that the synthesis of these proteins occurs on rough endoplasmic reticulum which fills the surrounding cytoplasm and that the proteins are subsequently transported to their site of storage. The soybean protein body is thus conceived as being a kind of vacuole. Its function is apparently repository in contrast to the protein bodies in maize which both synthesize and store the storage protein (Burr and Burr, this volume). Bar = 1 μm.

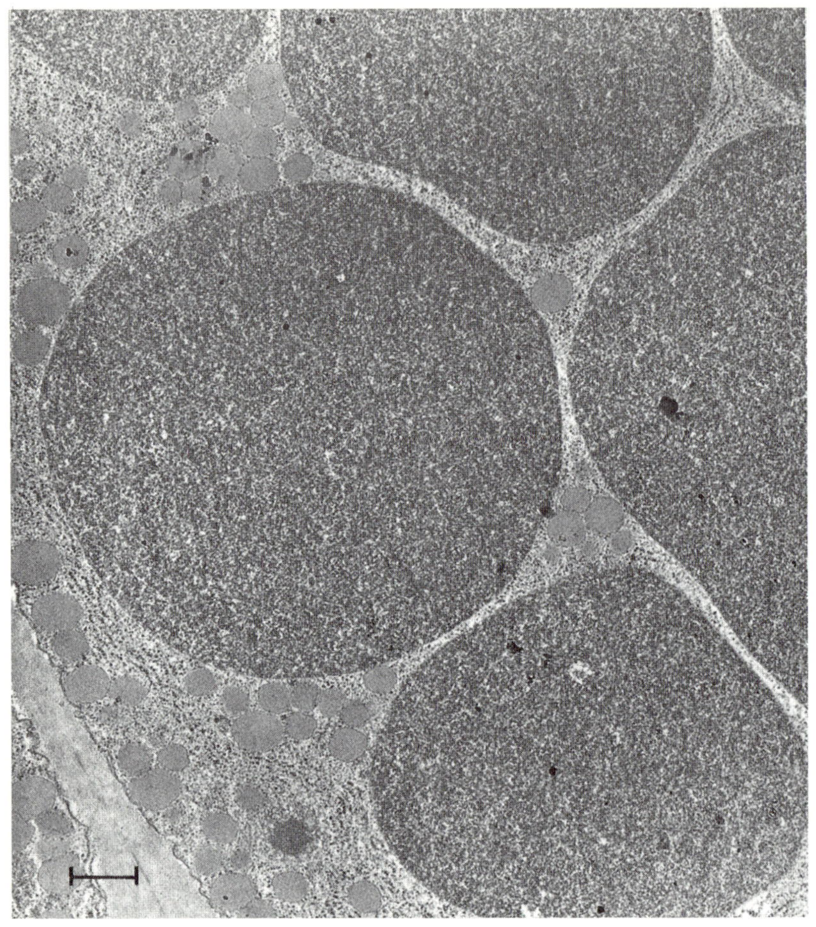

FIGURE 2. A later stage in the development of the protein body of soybean. The protein body vacuole is now completely filled by the accumulated storage proteins. Bar = 1μm.

REGULATORY VARIANT AND MUTANT ALLELES IN HIGHER ORGANISMS AND THEIR POSSIBLE ORIGIN VIA CHROMOSOMAL BREAKS[1]

Michael Freeling
James C. Woodman

Department of Genetics
University of California
Berkeley, California

The awesome mechanical complexities of a higher organism are, at least in large part, encoded in chromosomal DNA. It is now clear that only a small fraction of this DNA specifies the primary structure of particular polypeptides and the stable RNAs; the remainder of DNA sequences serve unknown functions. There is general agreement that some of this DNA must "regulate" individual structural genes and "program" development. Mutants in such program and regulatory DNA sequences could, in theory, help us traverse the hierarchies of organization that lie between the genome and the phenotype. The use of mutants to reduce one biological level to another is one definition of the genetic approach.

THE GENE IN HIGHER ORGANISMS

The most inclusive definition of a gene is the entire stretch of DNA involved in the controlled readout of a single product RNA. This stretch of *cis*-acting function is best defined by all possible transpositions that disconnect genic components. The traditional complementation group (Benzer, 1959), defined by pairwise combinations of all possible intragenic aberrations and nucleotide pair substitutions, should also delineate the largest possible gene. Unfortunately, dif-

[1]*Supported by USPHS, NIH, GM 21734. James C. Woodman is the recipient of a USPHS predoctoral grant NIH GM 07127.*

ISBN 0-12-602050-7

ferent gene components may require different types of lesions to display mutant behavior. Finally, it is possible that any gene component may be used in more than one gene simultaneously or that sequences may change orientation naturally (e.g., Wallace and Kass, 1976) or move in and out of genes (McClincock, 1950).

Figure 1 depicts a gene involved in producing a primary transcript which is processed to a message which is then translated into polypeptide. The structural gene specifies the amino acid sequence of the polypeptide. Other components may be transcribed as well, but are not translated or are removed during RNA processing.

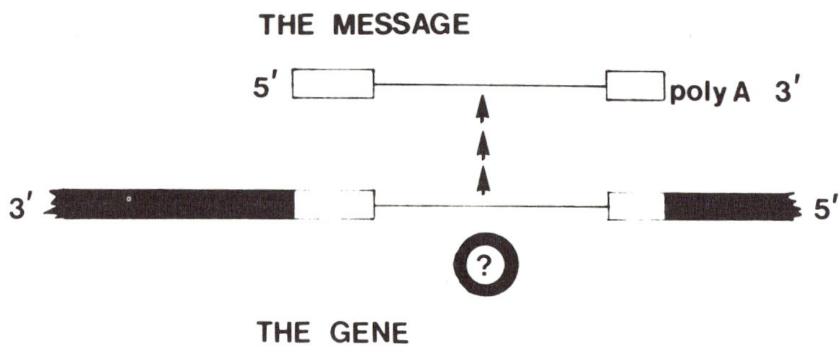

Cis-acting Regulatory Component (■ & □)

Structural Gene Component (——)

WHEN?

WHERE?

HOW MUCH?

HOW FAST?

WITH WHOM?

WHAT?

Figure 1. The gene in higher organisms. This idealized structure does not distinguish among cis-acting DNA (1) that which is involved in "autoregulation," (2) that which might be differentially programmable during development, and (3) that which might merely occupy space between functional genic components. A circle of DNA has been included near the structural gene component to indicate the possibility that sequences which do not encode polypeptide normally exist within coding DNA.

Most eucaryotic structural genes that have been analyzed at the DNA sequence level include extra DNA--introns-- which is not reflected in the final RNA transcript (Breathnach *et al.*, 1977; Jeffreys and Flavell, 1977b; Goodman *et al.*, 1977; Tonagawa *et al.*, 1977; White and Hogness, 1977; Tilghman *et al.*, 1978; Valenzuela *et al.*, 1978). In the case of a yeast tRNA gene, the insert DNA is transcribed but subsequently removed during processing (Knapp *et al.*, 1978). The generality and function for such disconnected structural genes are not known. As with the noncoding sequences in hnRNA, our knowledge is of structure only.

Structural genes probably comprise but a small percentage of the nuclear genome of any higher organism. We have calculated this percentage for maize, which has about 4 pg of DNA per unreplicated, haploid chromosomal complement (Bennett and Smith, 1976). Assuming that maize has 10,000 structural genes--twice *Drosophila melanogaster's* lethal complementation groups (see Garcia-Bellido and Ripoll, 1978)--and that each structural gene is 1300 nucleotide pairs long (NTP), structural genes comprise 1.3×10^7 NTP or 6.5×10^{-3}pg or 0.16% of the genome. Quantitative estimates of structural gene number based on polysomal RNA species complementary to "unique" genomic DNA per cell type or organ may run 20,000-50,000 structural genes per monoploid nucleus (see Goldberg *et al.*, 1978). In the case of maize, structural gene DNA would remain below 1% of the genome.

It is tempting to endow *cis*-acting regulatory regions with functions diagnostic for higher organisms. In particular, the classical embryological phenomena of determination and somatic heredity (Wilson, 1925; see Steeves and Sussex, 1972, for plants) must surely have explanations at the level of DNA, and perhaps have correlates at specific genes. While it is possible that individual structural genes are preset (as a part of determination) by changes involving *cis*-acting DNA--these changes being inherited somatically--it is also possible that structural genes generally react to developmentally important programming events at other levels of the genome (e.g., sensor and integrator levels of Britten and Davidson, 1969).

CRITERIA FOR "REGULATORY" ALLELES

Our discussion of "regulatory" alleles will be confined to alleles of genes that contain one structural gene component.

Regulatory mutants or variants are distinct from an arbitrarily chosen wild-type allele by displaying a difference in (1) regulatory-type behavior which (2) maps to the gene and which (3) is caused by a difference confined to nonstructural gene DNA. The mechanism underlying this cause might be pretranscriptional, transcriptional, RNA processing/turnover or translational. Naturally occurring variants in higher organisms often satisfy criteria (1) and (2) above, but rarely satisfy criterion (3). However, numerous variant alleles behave in ways difficult to explain at the structural gene level. On the other hand, induced mutants rarely exhibit regulatory behavior. It will be shown that chromosomal breaks seem to be implicated in the origin of those few induced mutants that do.

Studies on particular natural variant (not induced mutant) alleles will be cited in order to clarify the three criteria. We highlight some of our own results with the *Adh1* gene in maize.

By the early 1950s, molecular access to the gene appeared likely (see the Cold Spring Harbor Symposium in Quantitative Biology, Vol. 16, 1951). The phenomenological discoveries depicting a complexly regulated gene as a part of the chromosome (Goldschmidt, 1951) were to be overshadowed by the molecular biology of the structural gene component for about 25 years. Today, the combination of the few biochemically accessible specific gene systems, mutant recovery procedures, high resolution genetic mapping, and DNA cloning may permit a fresh approach to the venerable questions of classical genetics, identified by position effects, unstable genes, transposable elements, paramutation, and the like.

Criterion 1: Regulatory Behavior

Variants Involving the Timing and Cell-Specificity of Unstable Allele Reversion. The presence of anthocyanin pigmentation in maize aleurone and plant has led to a myriad of mutant or variant alleles affecting the expression of the several epistatic genes involved.

McClintock (1967) has reviewed the general topic of controlling element systems as they relate to programming individual genes during development. In general, one genetic element is always near or in the structural gene component, and usually inactivates or lowers its expression. In the presence and only in the presence of another element, the controlling element, the inactivated gene responds by a high frequency of mutation (reversion) to partial or full activity. The result is a genetic mosaic. Combinations of structural gene element and controlling element dictate precise cell specificities and division specificities for this mutator function. Further, either element may change in both positions in the genome, as evidenced by mapping, and timing-function, as evidenced by patterns in mosaic tissue. These mutational changes have been termed *transpositions* and *changes of state*, respectively. If a structural gene is, for some reason, routinely unstable, a controlling element system is not implied. In the absence of the *trans*-acting controlling element (mutator), the structural gene elements are stable. A comprehensive review exists on this subject (Fincham and Sastry, 1974), and Peterson (1976) has been especially careful with terminology.

Some controlling element systems were discovered in nature, like *Dotted* (Rhoades, 1938) and variegated pericarp, P^{VV} (Barclay and Brink, 1954). However, most alleles studied were induced, directly or indirectly, by chromosome breaks (McClintock, 1950, 1951). These will be discussed as "mutants" although the distinction between mutants and variants may not prove meaningful when chromosome breaks per se are the mutagen.

Some of the most intriguing results on the potential of alleles to dictate their own intensity or pattern of expression comes from studies at the *R* anthocyanin-conditioning locus in maize. Specifically, the phenomenon of paramutation was discovered and has been most thoroughly studied using variant alleles (Brink, 1973). *Paramutation* is an interaction between alleles whereby one or both are always or nearly always changed; and these changes are heritable. At *R*, the phenotype monitored is the timing of anthocyanin pigmentation during aleurone and plant development--as deduced from the size of pigmented clones--and also the positions of variegation. A mechanism for paramutation has been proposed that involves the over- or underreplication of negative, quantitative control elements ("metameres") acting in *cis* to hypothetical *R* structural genes (Brink, 1964; Sastry, Cooper and Brink, 1965). Paramutation has not yet been identified at any simple gene that specifies an accessible polypeptide product. However, ribosomal DNA magnification in *Drosophila* displays paramutational properties (Tartof, 1971).

Over- or Underproduction of a Specific Polypeptide. Genes that clearly contain a structural gene component(s), as deduced from stable RNA or polypeptide variants, present the possibility of eventually localizing a quantitative site within the locus. Chovnick and coworkers (review: Chovnick *et al.*, 1977) were able to map a *cis*-acting quantitative site outside of the xanthine dehydrogenase structural gene. In this case, the variant *rosy* allele specified differences in total number of XDH polypeptides per *Drosophila*. In the absence of the conditions necessary for gene fine structure determinations, similar arguments for variants affecting the rate of structural gene component expression from without have been indirect (Swank *et al.*, 1973, for the glucuronidase gene in mice, and thalassemias involving the α and β globins; Kabat and Koler, 1975). Often, variation in quantity of polypeptide per organism or per cell has been shown to involve timing of gene action during development (Schwartz, 1962, for pH 7.5 esterase in maize; Dickinson, 1975, for aldehyde oxidase in *Drosophila;* Boubelik *et al.*, 1975, for mouse H-2 antigen attachment to membrane; and Torres, 1974, for *Adh1* in sunflower). With the exception of the human thalassemias, none of the genes noted above has been assayed at the level of mRNA/cell. Even in the case of the globin messages, transcription has not been distinguished from RNA processing.

The *alcohol dehydrogenase-1* gene *(Adh1)* in maize is apparently regulated by a compensatory mechanism (Schwartz, 1971); *Adh1* variants (Schwartz, 1966) differ in their balances of allozyme expression. These variant alleles are electrophoretically distinguishable, as well as altered in hypothetical *cis*-acting components. Schwartz (1971) has explained his results by a "gene competition" hypothesis, in which a pair of alleles do not limit their own expression, but rather compete for a limited factor that is specified elsewhere in the genome. Birchler (1978) has proposed an alternative, but more complex, explanation. Freeling (1975) has used variants at *Adh1* to demonstrate that both unlinked *Adh* genes, *Adh1* and *Adh2,* are compensatorily regulated during anaerobiosis. Variants of compensatory-type phenomena have also been reported in *Chironomus* (Thompson and Patel, 1972). Although the regulatory-type behavior exhibited by the *Adh1* variants is not readily explained by differences in their allozyme subunits, such explanations have not been rigorously excluded.

Induction-Repression Variants. With the possible exception of a mouse β-glucuronidase allele (Dofuku *et al.*, 1971), variants defining correlates to bacterial *i, o,* and *p* genes are conspicuously lacking. It is also true that, with the exception of plant ADHs, inducible enzyme systems in higher organisms have not been studied genetically.

Circuitry Alleles. Alleles of simple genes that differ in their on-off circuitry (e.g., off in Malpighian tubules but on in fat body cells; on in pollen but off in endosperm) have not been found. Alleles at *R* in maize appeared to be of this type, but a duplicate locus was shown to be involved (Brink, 1973); several of the complex (multicistronic) morphological loci in *Drosophila,* especially those with some dominant alleles, display much allelic diversity (e.g., *Notch, bithorax).*

Organ-Specific Reciprocal Effect at Maize Adh1. Measures of allelic diversity are traditionally confined to structural gene components via allozyme analyses. Woodman has tested *Adh1* in 20 maize cultivars representing at least 6 races of maize. Cultivars were usually homozygous for *Adh1,* as assayed by electrophoretic mobility. *Adh1-S* alleles specify an ADH subunit conferring a relatively slow electrophoretic mobility to ADH·ADH dimers; *Adh1-F* specifies an electrophoretically faster product. Each *Adh1-F* cultivar was crossed to the inbred line carrying the *Adh1-S* standard allele, and each *Adh1-S* cultivar was crossed to the inbred line carrying the *Adh1-F* standard allele. The resulting *Adh1-S/Adh1-F* heterozygote seeds, plants, and pollen were analyzed electrophoretically for the *balance* of ADH1-S versus ADH1-F subunit expression (Woodman, 1978). Figure 2 shows some representative electrophoretograms. Much variability in allozyme balance was found. These balances are allelic to *Adh1* (see next section). Most importantly, heterozygotes which underexpress one allele in the scutellum overexpress that same allele in the root, and *vice versa*. A reciprocal relationship exists for *Adh1* expression between maize organs. It is difficult, but never impossible, to imagine structural gene component variation accounting for this organ-specific reciprocal behavior.

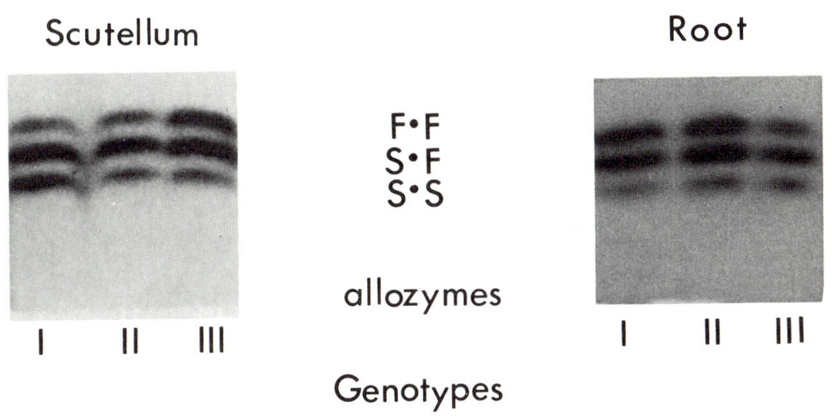

FIGURE 2. Electrophoretograms displaying the organ-specific reciprocal effect at Adh1. *All samples are heterozygotes for* Adh1 *alleles which specify ADH subunits which are either "slow" or "fast" in anodal electrophoretic migration rates. Three genotypes are shown: I, which gives a relatively high ADH1-S expression in scutella but relatively low ADH1-S in anaerobically induced primary roots; II, which is intermediate in allozyme ratio; and III, which expresses ADH1-F subunits preferentially in the scutella but ADH1-S in the root. The dimeric subunit structure of ADH is diagrammed. Only ADH1·ADH1 (Set I) isozymes are shown; in the anaerobically induced root, other ADH isozymes also occur (Freeling, 1973).*

Criterion 2: The Regulatory-Type Difference is Allelic to the Structural Component

Mutant alleles with recessive phenotype are expected to segregate away from a marker allele during meiosis and to fail to complement other recessive alleles in mutant/mutant heterozygotes. If a presumptive regulatory allele is either paramutable or paramutagenic (Brink, 1973), complications of this classical allelism test may result. A locus defined by dominant variants or mutants, such as certain dwarfisms or morphological aberrations, presents special difficulties for allelism tests since a heterozygote would probably express both nonwild phenotypes; very close linkage based on low recombination frequencies (10^{-4} in maize; 10^{-5} in *Drosophila)* would support allelism but not prove it (see Freeling, 1978, for difficulties in interpreting intragenic recombination frequency results). Of course, without an assay for a gene's RNA or polypeptide product, there is no way to resolve the structural gene component from other genic components.

When the protein product of a gene is accessible, allozymes are valuable tools in testing regulatory-type lesions for allelism. Figure 3 graphically represents the sort of data that we have used to prove that certain *Adhl* variants--identified originally on the basis of an atypical allozyme ratio (Fig. 2) --are allelic with, and act in *cis* to, the *Adhl* structural gene--identified by an electrophoretic mobility site difference. The alleles being compared in Fig. 3 are *Adhl-F* (Standard) and *Adhl-33F* (from Super Gold Pop cultivar); their products are probably identical and have the same electrophoretic mobility. An *Adhl-F/Adhl-33F* heterozygote was crossed to an *Adhl-S/Adhl-S* tester; all progeny generate three allozymes. However, when the *ratio* of these allozymes is carefully recorded, these testcrossed progeny scutella showed a bimodal distribution (modes at intervals 27-28 and 32-33% *F·F* homodimer of total ADH activity). When roots and pollen of these same individuals were analyzed, bimodality was even more extreme, but the roots showed the reciprocal effect mentioned previously. When plants with a given % *F·F* in scutella were self-pollinated, the 50% heterozygotic progeny showed the allozyme balance characteristic of their parents. Histogram C in Fig. 3 gives four examples of these data. It was proved that the "balance" gene segregating in Fig. 3 is allelic to the *Adhl* structural gene by backcrossing *Adhl-F(or 33F)/Adhl-S x Adhl-S/Adhl-S*. In all cases, the heterozygotes displayed only one allozyme ratio with no evidence (1) for segregation of the "reciprocal effect" site away from the electrophoretic site; (2) for *trans* effects on *Adhl-S* expression; or (3) for paramutagenic effects which might be expressed by *Adhl-S* in subsequent generations (Woodman, unpublished data). Note that allelism and *cis*-action

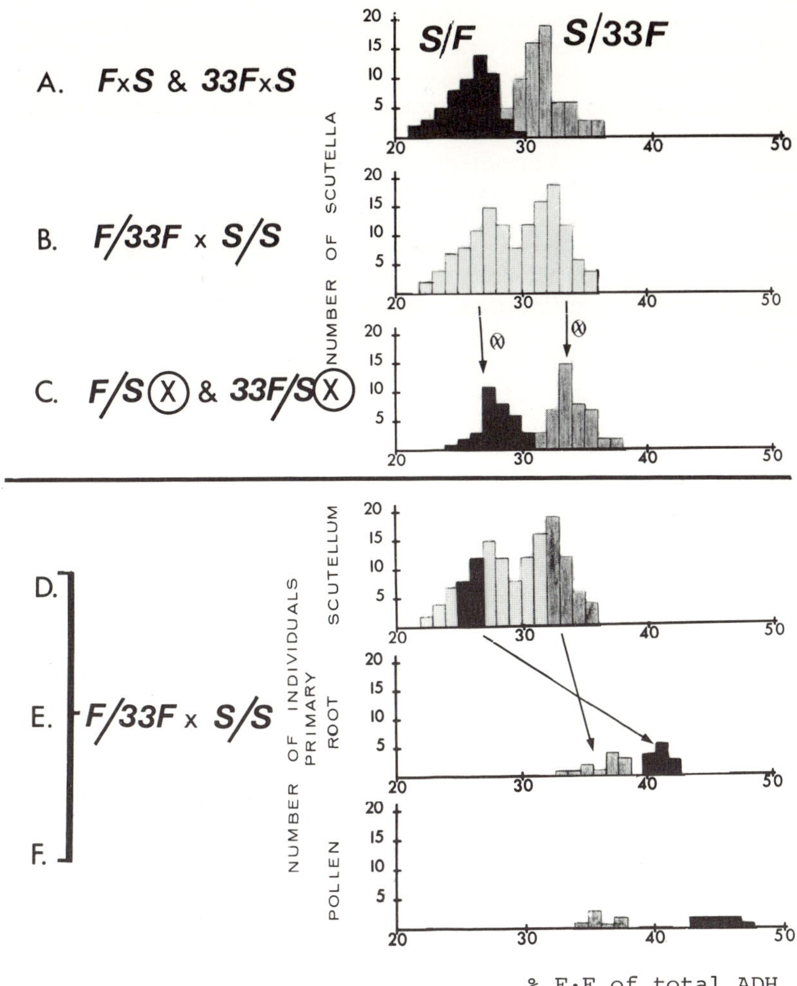

FIGURE 3. Histograms plotting number of individuals in relation to their ADH1·ADH1 allozyme (Set I) ratio; see Fig. 2 for typical electrophoretograms. Histograms A-C, inclusive, depict the (A) parental % F·F data, (B) testcross data, and (C) F_2 *data. Histograms D-F, inclusive, demonstrate the reciprocal effect in three "organs" of individual test-crossed progeny. Black areas are known or deduced to be* F/S *in genotype; lightly shaded areas are known or deduced to be* 33F/S*. The bimodal distributions shown in identical histograms B and D are clearly the overlap of* S/F *and* 33S/F *distributions.* S *is the abbreviation for the allele* Adh1-S*, etc. Allozyme activity ratios were quantified by controlled densitometric scans of starch gels which had been stained for ADH activity.*

have been demonstrated without use of flanking markers. When dealing with regulatory gene components and subtle quantitative variants that may be extremely polymorphic, the introduction of close flanking markers is not necessary and not always desirable.

Criterion 3: The Allelic Regulatory-Type Difference is Caused by a Lesion Confined to Nonstructural Gene DNA.

Ideally, one would compare the DNA nucleotide sequence comprising the entire wild-type gene and the regulatory mutant derivative; the structural gene would be identified in the genomic DNA fragment using messenger RNA or DNA complementary to it (e.g., R-looping). Such a straightforward comparison has not yet been made for any regulatory mutants or variants in higher organisms, but this sort of data will probably be available soon for human thalassemia variants (Jeffreys and Flavell, 1977a). This direct approach is greatly facilitated by recent advances in molecular cloning technology.

Comparisons among polypeptide products or enzyme activities specified by regulatory mutants and the wild-type progenitor allele are more easily accomplished. Enzyme kinetic parameters, rate of thermal inactivation, rate of subunit dissociation-reassociation, electrophoretic mobility, enzyme subunit conformation, various measures of immunological cross-activity, peptide "fingerprint," and partial or complete amino-acid sequence may be measured. If the regulatory mutant lesion is confined to nonstructural gene DNA, the protein product should remain unaltered. Since enzymes may be processed from larger polypeptides and since different DNA sequences may specify identical polypeptides, even a comparison at the level of total amino acid sequence is not unequivocal proof of structural gene identity.

Genetic fine structure mapping is another tool by which a lesion may be unequivocally assigned to *cis*-acting DNA. To do this, one must first map the structural gene component using mutants that clearly alter the enzyme protein (e.g., using electrophoretic mobility, or intragenic complementation).

There must be a convincing argument that the extremes of the structural gene have been approached. The presumptive regulatory lesion may then be mapped well outside of the structural gene component, as measured in map units or coconversion frequencies. This has been clearly accomplished by microbial geneticists and by Chovnick and co-workers in *Drosophila's rosy* gene (Chovnick *et al.*, 1977). The disadvantage of the genetic fine structure proof is its dependence on low-frequency genetic recombination. Not only is there no guarantee that variant wild-type alleles will recombine intragenically (Freeling, 1978), but intragenic chromosomal aberrations--such as intervening sequences--might be expected to interfere with heteroduplex formation.

SOME EVIDENCE ON MUTAGEN SPECIFICITIES TO GENE COMPONENTS

Only a few genes in higher organisms presently have the mutant selection procedures, genetic resolution, and accessibility to gene product necessary to assess (1) the specificity of a mutagen for different types of DNA, or (2) the specificity for different gene components in expressing similar DNA lesions. *Rosy* in *Drosophila,* which encodes XDH polypeptides, and the *Adh* gene systems in maize and *Drosophila* are examples.

Chovnick *et al.*, (1977) have probably saturated the *rosy (ry)* structural gene component with XDH-deficient mutants (rosy eye color) and XDH-low mutants (purine sensitive but normal eye color) using X rays and ethyl methanesulfonate (EMS). All of the over 100 intragenic mutants recovered behaved as points recombinationally, and have been located at or between the right and left boundaries of the structural gene component. These boundaries were defined using variant sites conferring electrophoretic mobility differences to XDH. It was concluded that the *cis*-acting regulatory component(s) of *ry* was either very small, or immutable, or that point mutations within it were not recognizable. Subsequent work with variants (not mutants) did identify an allelic regulatory component mapping well outside of the structural gene, which we have discussed previously.

Chovnick has preliminary evidence for EMS-induced regulatory alleles affecting levels of XDH (Chovnick, personal communication), but very low or dysfunctional mutants are still lacking.

Schwartz (e.g., 1969) has used EMS to induce mutants of the *Adh1* in maize. His protocol is unique in that mutants are recognized and selected solely on the basis of altered allozyme patterns; the electrophoretic protocol we use for *Adh1* mutant classification (subsequent section, Fig. 5) is based on

Schwartz's methods. Well over 100 EMS induced mutants of *Adh1* have been studied extensively by Schwartz and co-workers; although several appeared to be low producers, all are most simply explained by lesions in the structural gene (Schwartz, personal communication). Twenty-one of these mutants have been tested for ability to recombine intragenically and all behave as unique points (Freeling, 1976, 1978). The absence of presumptive regulatory mutants (e.g., over- or underproducers with unchanged product ADH, changes in organ-specific expression) among the EMS-induced mutants is particularly meaningful since several *Adh1* variants and some of our radiation-induced mutants display obvious regulatory-type behavior, as will be discussed.

Most likely, the regulatory functions of *cis*-acting components are difficult to mutate by point lesions. This would be expected if some of the allelic regulatory and program DNA sequences were redundant, as was hypothesized by Britten and Davidson (1969).

REGULATORY MUTANTS AND THEIR ORIGINS

The list of candidates (satisfying criteria 1 and 2) for regulatory mutants (not variants) in higher organisms is short indeed. Since Beadle and Tatum's (1941) results from which the rule "one gene-one polypeptide" evolved, volumes have been written on molecular mutagenesis. It is interesting that the theory of the gene has not progressed far from Goldschmidt's (1951) account.

One might be tempted to look to lower eucaryotes for a gene somehow intermediate in complexity between the well-studied bacterial complementation group--such as *lac*, *gal*, and *tryp*--and the gene in higher organisms. At present, it is unclear whether or not organisms such as *Aspergillus*, *Neurospora*, or the yeasts truly present such an intermediate situation. The bacterial genes that have been most thoroughly studied possess *cis*-acting regulatory sites, promotors, and operators, which involve their response to small molecules: inducer specificity, ability to induce, ability to repress, sensitivity to catabolite or aminoacyl tRNA repression. The general importance of autoregulation, the sensitivity of a gene to its products and the resulting regulatory loops, has been reviewed recently (Calhoun and Hatfield, 1976). Even at its most complex, the inducible bacterial genes that have been studied seem to *respond* immediately to their environment. Since few genes in higher organisms are inducible in the microbial sense, it has been difficult to extend the concept of autoregulation. Genes in higher organisms are clearly *programmed to behave differently in different cells, positions, or developmental times;* the microbial counterparts to programmable gene components are lacking. With the possible exception of some interesting results on the genetics of mating type in yeast (Herskowitz *et al.*, 1977), we do not know of any evidence for *programmable* gene components in lower eucaryotes; all mutants are in induction-repression functions. Haynes (1975) clearly separated a *cis*-acting superinduction site from the *acetamidase* structural gene in *Aspergillus nidulans*. A possible low-producer site for repressible alkaline phosphatase in *Neurospora crassa* has been reported (Gleason and Metzenberg, 1974; Lehman and Metzenberg, 1976). Constitutive mutants in nitrite reductase (Rand and Arst, 1977) and uric acid permease (Arst and Scazzocchio, 1975) have been shown to be *cis*-acting. Mutants conferring nonrepressibility to *alcohol dehydrogenase-2* (Ciriacy, 1976) and *carbamylphosphate synthetase-1* (Thuriax *et al.*,1972) have been reported in baker's yeast. A site conferring insensitivity to carbon catabolite repression maps between its two target proline catabolism genes (Arst and MacDonald, 1975). Lukaszkiewicz and Paszewski (1976) reported on a mutant sulphate permease gene in *Aspergillus* which behaved like a hyperrepressible operator. In summary, sohpisticated genetic analyses of genes in lower eucaryotes have not as yet discovered concepts beyond those already known to operate in bacteria. In addition, those regulatory mutants that have been studied were, by and large, spontaneous or induced under conditions where base substitutions might be expected. It is not apparent from the literature that the developmental complexities of lower eucaryotes are being exploited in regulatory mutant searches.

General aspects of controlling element systems have been discussed in the previous section under "variants". McClintock (1950, 1951) "induced" new alleles of activator-dissociator *(Ac-Ds)* and suppressor-mutator *(Spm)* using a chromosomal breakage-fusion-bridge (BFB) cycle. A recent review (Fincham and Sastry, 1974) diagrams and explains clearly both chromatid and chromosome BFBs. In general, certain chromosomal aberrations lead to crossover events which generate broken ends. In the gametophytes (n) and endosperm (3n), broken ends may replicate their ability to subsequently rejoin, permitting a chromatid BFB. Chromosomal BFBs may occur anywhere in the plant. BFBs are used to force the plant regularly to undergo chromosome snaps at mitotic anaphase. Since controlling element systems almost certainly involve inserted DNA sequences, it was suggested that chromosome snapping per se may liberate inserts; McClintock (1951) has shown clearly that BFBs involving chromosome 9 lead to both receptor (like *Ds*) and regulatory (like *Ac*) elements on other chromosomes.

The origin of controlling element systems has been most thoroughly studied using *Dotted (Dt)*. *A* is one of the genes specifying anthocyanin pigmentation in the plant and aleurone. The most common recessive allele, *a*, was thought to be a stable variant; the colorless aleurone almost never showed anthocyanin-positive sectors which would mark an *a*→*A* reversion. In the presence of a variant gene, *Dt*, at about 0 position on chromosome 9, *a* became unstable (Rhoades, 1938). The phenotype was a colorless seed covered with small, pigmented dots indicating reversional events late in endospermal development. Neuffer (1955) found two new *Dt* alleles as variants in South American Indian corn; these mapped to chromosomes 6 and 7.

McClintock (1950, 1951) attempted to induce new *Dt* alleles by establishing chromosome 9 BFB in *a/a, dt/dt* lines: 117 dotted seeds occurred among 93,078. None transmitted, so unequivocal tests for *a* involvement were not possible. However, it was intriguing that chromosome breaks generate instability (*Dt*) where x rays applied to immature kernels 73-81 hours postpollination proved nonmutagenic in about 9 x 10^5 tries (Stadler, 1944). Doerschug (1973) examined 1.5 x 10^5 seeds of *a/a, dt/dt* resulting from pollen parents undergoing chromatid BFB. Of the 250 seeds that showed some aleurone dots, two new *Dt*s were recovered: One was like the original *Dt* and mapped close to the telomere on the short arm of chromosome 9; the other was unlinked to chromosome 9 and also showed a new mosaic pattern ("state") where *a*→*A* reversion events are concentrated at the crown of the kernel.

The *cherry* allele of the *white(w)* locus in *Drosophila* was recovered following x irradiation (Green, 1967). This allele is interesting because it is unstable; changes to dysfunction (white eye) were sometimes mediated by deletions with origins at *w*, and transpositions of parts of *w* from X to chromosome 3 have been reported (Green, 1969). Green (1977a) has made a compelling case for the insertional nature of unstable mutants in *Drosophila*.In maize, Mottinger (1970, 1973) made yet another attempt at inducing intragenic mutants in male germ cells using x rays. With a few exceptions (cf. Nelson, 1968; and see section on maize *Adh1*), x rays have not generated lesions confined to a gene, although chromosomal aberrations abound. Mottinger found no point mutants in the *Sh-Bz2* region of the short arm of chromosome 9, although deletions occurred. Premeiotic x ray treatment did induce some unstable *bronze-2 (bz2)* mutants.

Male recombination (*MR*) chromosomes in *Drosophila* have at least three effects on flies that carry them: heightened frequencies of mitotic recombination in males (Hiraizumi, 1971), lowered transmission of the chromosome carrying *MR* factors, and greatly accelerated mutant frequencies for some but not all genes tested (Green, 1977b). These unstable mutants display characteristics that Green ascribes to an insertional mutant. Further, *MR* chromosome IIs may have paramutator qualities (Matthews *et al.*, 1978).

Position effects in *Drosophila*, obviously breakpoint-associated, have been reviewed (Baker, 1968).

Chromosome breakage may cause or mediate special types of mutants, including gene instabilities.

NEW MUTANTS IN MAIZE *Adh1* BY POLLEN SELECTION AND CHARACTERIZATION BY ALLOZYME BALANCE

The *Adh1* gene in maize presents advantages for mutational analyses (Freeling and Cheng, 1978 and citations therein). Foremost among these is the ability to select chemically rare ADH-negative or ADH-low pollen grains owing to their resistance to allyl alcohol vapor. ADH specified exclusively by the *Adh1* gene is synthesized in the pollen after meiosis. For this reason, ADH levels per gametophyte reflect the haploid *Adh1* genotype; Mendelian segregation may be visualized by staining pollen for the presence or absence of ADH, as shown in Fig. 4. Therefore, if an *Adh1-dysfunction* mutant were induced at or before meiosis, all gametophytic nuclei--the synthetically active tube nucleus and both sperm--should be *Adh1*$^-$ in an ADH$^-$ pollen grain. If the *Adh1*$^-$ mutant is not accompanied by lethal deficiencies, it should confer allyl alcohol resistance and transmit through the male. In order to distinguish the target (male) allele from the *Adh1* of the tester (female), electrophoretically distinguishable *Adh1* variants were used, as shown in Fig. 5. Unlinked genetic markers are also employed. The choice of *Adh1-S* as a target allele was based on one of our (MF) previous studies on intragenic recombination and spontaneous revertant/forward mutant frequencies.

Given the argument that chromosome breakage events were more likely than point mutations to generate regulatory alleles at *Adh1*, we used high-energy neon ions accelerated to 400 MeV/amu at the Bevatron, Lawrence Berkeley Laboratory. 250 KV x rays and mutants arising spontaneously were used as controls. Sporocytes, average maturity of prophase I, were used as targets. Pollen was then selected with levels of allyl alcohol permitting escape of some *Adh1*$^+$ gametophytes and was crossed onto tester ears. We hoped, perhaps naively, that heavy-ions would generate multihit chromosomal breakage events with single-hit kinetics; perhaps redundant, *cis*-acting *Adh1* sequences would be altered.

Sixty-nine *Adh1 - altered* mutants have been recovered as single, allyl alcohol resistant gametophytes following ionizing radiation treatment of the immature tassel; the median micro-

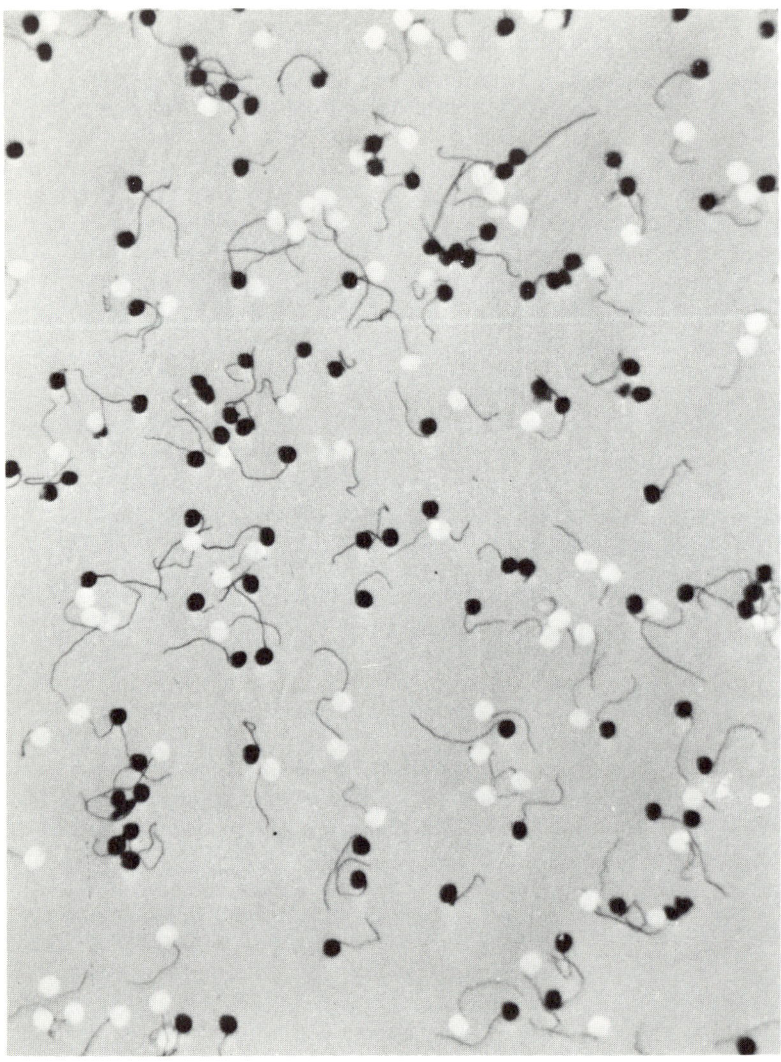

FIGURE 4. $Adh1^+/Adh1^-$ *pollen which was germinated on media before cytochemical staining for the presence of ADH activity. The dark gametophytes are* ADH^+.

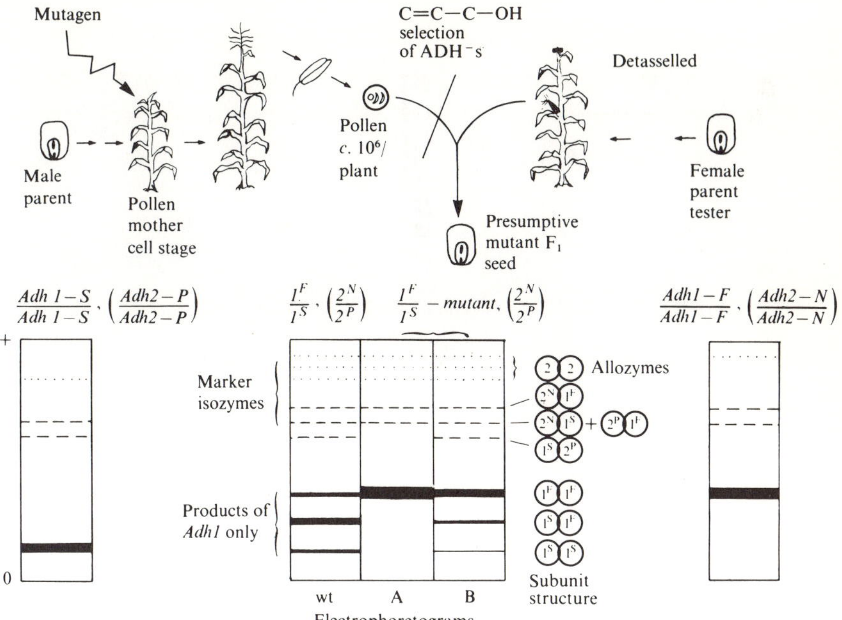

FIGURE 5. Chemical selection, recovery and classification of Adhl-*mutants. The paternal or target* Adhl *allele,* ADHl-S *is abbreviated* 1^S*, etc. The electrophoretograms depict semi-quantitative banding patterns and intensities after native starch gel electrophoresis. 0 denotes origin; + denotes the anode. The ADH activities reflected in these diagrams are from the wild-type* F_1*(wt),* Adhl-negative *heterozygote (A) and* Adhl-S underproducer *heterozygote (B). Extracts were obtained from scutellar slivers. The subunit composition(s) of each band is indicated (Freeling and Schwartz, 1973) where subunit* 1^S *is specified by* Adhl-S, *etc. The dashed and dotted bands in the marker isozyme region of gel denote low activity and usually not present, respectively (Courtesy of Cambridge University Press, Freeling and Cheng, 1978).*

sporocyte was in meiotic prophase I. Individual mutant heterozygote seeds were placed into one of four classes on the basis of scutellar allozyme pattern following electrophoresis and ADH staining. Figure 6 depicts representatives of these four classes: (A) dysfunction, (B) underproducer, (C) overproducer, and (D) up-ADH_2 activity. Mutants have been recovered and confirmed in the first three categories, but all of the 22 Ne^{10+}-induced dysfunctions did not transmit past the F_1. Conclusions were drawn from our results (Freeling and Cheng, 1978): Ionizing radiation at meiosis did not cause any (0/69) lesions confined to the *Adhl* structural gene component; the

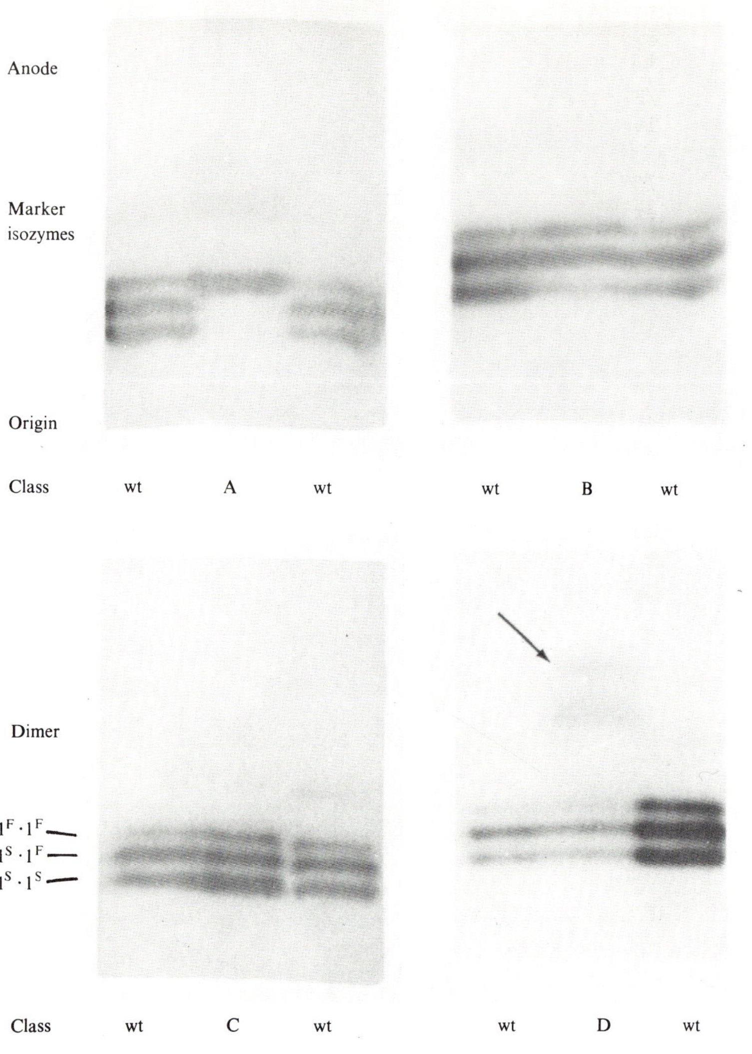

FIGURE 6. Four classes of Adh *mutants detected in* F_1 *seeds by abnormal allozyme ratios. Each electrophoretogram analyzes three extracts, each reflecting a single scutellar sliver. The nonmutant profile is designated "wt." Class A is a total* Adh1-S *dysfunction; class B is an allozyme balance skewed toward* Adh1-F *product; class C is an allozyme ratio skewed toward* Adh1-S *product; and class D includes apparent elevation of* Adh2 *gene products. The arrow marks the "fuzzy" ADH2·ADH2 isozyme, which is actually composed of three bands (see Fig. 5). (from Freeling and Cheng, 1978; courtesy of Cambridge University Press.*

mutants that did transmit were "unstable" in *Adh1-S* expression. Figure 7 shows individual backcross progeny of one of the class mutants (Ne^{10+}-underproducer, *S1945)*; the cross was *Adh1-F/S1945* x **F/F**. These derivatives are being analyzed. The next paragraph gives preliminary results on one of these mutant derivatives.

Adh1-S9151a is our first stable balance mutant; the "*a*" denotes a derivative of the original aberrant. This mutant is

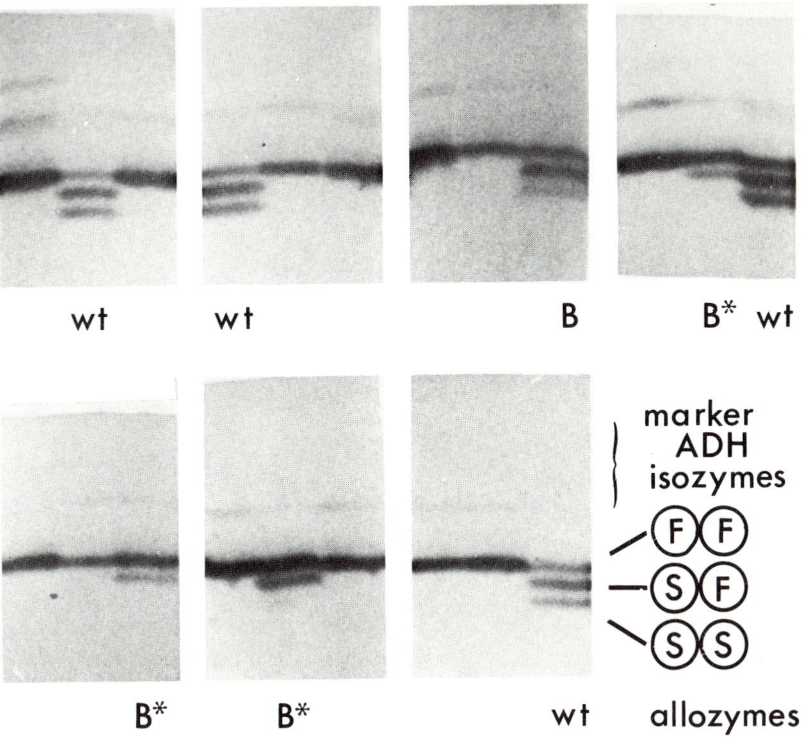

FIGURE 7. Instability of allozyme ratios among 21 backcross progeny of the cross Adh1-F/Adh1-F *x* Adh1-F/Adh1-S-mutant. *This mutant,* S1945, *was first recovered as an underproducer in the scutellum; these electrophoretic profiles are from scutellar slivers. The mutant allele was transmitted 8 times out of 21: "wt" denotes normal* Adh1-S *expression; "B" denotes class B underproducer; and B* denotes extreme underproduction of ADH1-S subunit activity.*

now a stable "underproducer" (class B) linked closely with an electrophoretic site in the *Adh1-S* structural gene. Its phenotype includes underproduction of ADH-S subunits in the scutellum, and--reciprocally--overproduction of ADH1-S subunits in the anerobic primary root. The mutant has no *trans* effects, as judged from experimental data similar to that used to prove allelism for reciprocal effect variants at *Adh1* (Fig. 3). The original F_1 seed, of the genotype *Adh1-F/Adh1-S1951*, was recognized as an underproducer in the scutellum, but this characteristic was not stable upon backcrossing to the *Adh1-F* parent (similar to data in Fig. 7). Electrophoretic analyses of scutellar slivers from backcrossed progeny (BC_1) found that most *Adh1-S* alleles appeared to be wild-type, one (of 75 transmissions) showed underproduction, several displayed overproduction, and some hyperploids for *Adh1* were transmitted through the F_1 male gametophytes and were detected in a test cross. We call these backcrossed individuals "derivatives." Meiotic chromosomes from some of these *Adh1-F/Adh1-S1951* derivatives have been examined; extra centric fragments are usually present, and chromosome 1 appears to be involved in a complex rearrangement that leads to some bridges at both anaphases. The single *Adh1-1951a* derivative (underproducer) was not analyzed and much further chromosomal work will be necessary if we decide that sorting out these complex aberrations is worthwhile. The underproducer derivative, *Adh1-S1951a*, was backcrossed to the *Adh1-F* parent once again (BC_2). Now the *Adh1-S1951a* mutant behaved as a stable allele of *Adh1* with a phenotype as described. Further, its chromosomes appear to be normal.

The "reciprocal effect" regulatory behavior of *Adh1-S1951a* would not be expected of a structural gene mutant. However, we have not unequivocally satisfied criterion 3; *Adh1-S1951a* is likely a regulatory mutant, but not proved to be one. The product of *Adh1-S1951a* appears nonmutant in electrophoretic mobility and thermostability (unpublished); a comparison at the peptide level is in progress.

The origin of *Adh1-S1951a* parallels that of maize transposable element systems in several ways, including an origin by chromosomally damaging treatment and an intermediate stage of gene instability.

As with the bulk of data dealing with radiation-induced single-gene mutants originating in the male germ line of higher plants, all of our original mutants were obviously associated with breakpoints. Freeling and Cheng (1978) have compiled and discussed much of these data; it was proposed that two-strand breaks are either not repaired or that extensive exonuclease action "rips out" several kilo-NTPs before reunion. If such a hypothesis proves true, our overall rationale for using accelerated high Z particles would be invalid.

Perhaps the most interesting of our results is that we were able to mimic a naturally occurring reciprocal effect variant with an induced mutant.

A HYPOTHESIS ON THE ORIGIN OF REGULATORY MUTANTS IN HIGHER ORGANISMS

In lieu of compelling data, we offer the following suggestions as to the nature of regulatory gene components and their mutability. Perhaps a considerable portion of a higher organism's complementation group (gene) is concerned with *cis*-acting regulation and perhaps small (<300 NTP) lesions in this region do not appreciably affect gene expression. Thus, point mutagens appear to confine their effects to structural gene components. We hypothesize that chromosomal breaks lead to regulatory mutations by (1) placing a structural gene within another control region, or (2) by deleting/duplicating quantitative DNA near a gene (see Brink, 1973, for concept of "metameres,") or (3) by otherwise rearranging (e.g., inverting, inserting new DNA) into *cis*-acting DNA. Our own data at *Adh1* certainly imply *quantitative, organ specific* gene control and may involve gene components that are differentially programmed during development. Perhaps more importantly, chromosome breaks per se may indirectly cause regulatory mutations by liberating insertional sequences which, in turn, integrate in or near any gene. McClintock (e.g., 1950, 1951) is certainly the originator of the "chromosome breaks per se" hypothesis. Our preliminary data on the origin of the reciprocal effect mutant *Adh1-S1951a* fit McClintock's view. We find it reassuring that the "chromosome breaks per se" hypothesis erases the meaningful distinction between naturally occurring alleles (variants) and induced alleles (mutants).

SOME FANCIFUL CONSEQUENCES OF THE HYPOTHESIS

The term progressive evolution might be defined as the earthly phenomenon of increasing organismal complexity over evolutionary time. To be more exact, if one looks at the most complex life form at time intervals from 3.5 billion years ago to the present, gradual change in the direction of some maximum of complexity is clear; human brain function is the present apex of complexity. Since natural selection depends ultimately on new mutations, it would be satisfying to understand the sort of DNA lesions that might have selective value. Base substitutions in structural genes seem unlikely candidates for "progressive" mutants. The hypothesis presented would favor chromosomal aberrations over base substitutions. It is interesting that Muller (1955, p. 155) held precisely the opposite point of view.

Those interested in breeding using induced mutants might fare better by utilizing insertional mutation systems or irradiating with high LET particles as opposed to using point mutagens (e.g., ethyl methanesulfonate).

Evoking chromosomal breaks as the first of several steps toward malignant transformation seems to fit much data on the mutational causes of cancer.

Speculation aside, we know nearly nothing about nonstructural gene DNA and its relationship to gene programming during development or autoregulation. Our ignorance should come as no surprise: Neither do we understand the genetic reason why any two bacterial taxa differ. Particular mutants that behave in ways which are, *a priori,* relevant to understanding one of life's mysteries *and* whose DNA may be studied directly at the level of base sequence should prove useful. Fortunately, a mutant's value is little affected by the obvious inadequacies of our own cleverness.

ACKNOWLEDGMENTS

David S. K. Cheng and Deverie Bongard gave excellent technical assistance. Roy Sahara tended our greenhouses and field. We also thank the Bevatron crew at Lawrence Berkeley Laboratory. We thank Virginia Walbot for helpful comments.

REFERENCES

Arst, H. N., Jr., and MacDonald, D. W. (1975). A gene cluster in *Aspergillus nidulans* with an internally located *cis*-acting regulatory region. *Nature 254,* 26-31.

Arst, H. N., Jr., and Scazzocchio, C. (1975). Initiator constitutive mutation with an "up-promoter" effect in *Aspergillus nidulans. Nature 254,* 31-34.

Baker, W. K. (1968). Position-effect variegation. *Adv. Genet. 14,* 133-169.

Barclay, P. C., and Brink, R. A. (1954). The relation between modulator and activator in maize. *Proc. Nat. Acad. Sci. (U.S.) 40,* 1118-1126.

Beadle, G. W., and Tatum, E. L. (1941). Genetic control of biochemical reactions in *Neurospora. Proc. Nat. Acad. Sci. (U.S.) 27,* 499-506.

Bennett, M. D., and Smith, J. B. (1976). Nuclear DNA amounts in Angiosperms. *Phil. Trans. Roy. Soc. London 274,* 228-274.

Benzer, S. (1959). On the topology of the genetic fine structure. *Proc. Nat. Acad. Sci. (U.S.) 45,* 1607-1620.

Birchler, J. A. (1978). On the nature of the compensation of alcohol dehydrogenase in a dosage series of 1L. *Maize Genet. Coop. News Lett. 52,* 30-32.

Boubelík, M., Lengerová, A., Bailey, D. W., and Matoušek, V. (1975). A model for genetic analysis of programmed gene expression as reflected in the development of membrane antigens. *Develop. Biol. 47,* 206-214.

Breathnach, R., Mandel, J. L., and Chambon, P. (1977). Ovalbumin gene is split in chicken DNA. *Nature 270,* 314-319.

Brink, R. A. (1964). Genetic expression of *R* action in maize. In *Role of the Chromosomes in Development.* (M. Locke, ed.), pp. 183-230. Academic Press, New York.

Brink, R. A. (1973). Paramutation. *Ann. Rev. Genet. 7,* 129-152.

Britten, R. J., and Davidson, E. H. (1969). Gene regulation for higher cells: A theory. *Science 165,* 349-357.

Calhoun, D. H., and Hatfield, G. W. (1976). Autoregulation of gene expression. *Ann. Rev. Microbiol. 30,* 275-299.

Chovnick, A., Gelbart, W. and McCarron, M. (1977). Organization of the Rosy Locus in *Drosophila melanogaster. Cell 11,* 1-10.

Ciriacy, M. (1976). *Cis*-dominant regulatory mutations affecting the formation of glucose-repressible alcohol dehydrogenase (ADHII) in *Saccharomyces cerevisiae. Molec. Gen. Genet. 145,* 327-333.

Dickinson, W. J. (1975). A genetic locus affecting the developmental expression of an enzyme in *Drosophila melanogaster. Devel. Biol. 42,* 131-140.

Doerschug, E. B. (1973). Studies on Dotted, a regulatory element in maize. Phase variation of Dotted. *Theoret. Appl. Genet. 43,* 182-189.

Dofuku, R., Tettenbron, U., and Ohno, S. (1971). Further characterization of o^s mutation of mouse β-glucuronidase locus. *Nature New Biol. 234,* 259-261.

Fincham, J. R. S., and Sastry, G. R. K. (1974). Controlling elements in maize. *Ann. Rev. Genet. 8,* 15-50.

Freeling, M. (1973). Simultaneous induction by anaerobiosis or 2, 4-D of multiple enzymes specified by two unlinked genes: Differential *Adh1-Adh2* expression in maize. *Molec. Gen. Genet. 127,* 215-227.

Freeling, M. (1975). Further studies on the balance between *Adh1* and *Adh2* in maize: Gene competitive programs. *Genetics 81,* 641-654.

Freeling, M. (1976). Intragenic recombination in maize: Pollen analysis methods and the effect of parental *Adh1*$^+$ isoalleles. *Genetics 83*, 701-717.

Freeling, M. (1978). Allelic variation at the level of intragenic recombination. *Genetics 89*, 211-224.

Freeling, M., and Cheng, D. S. K. (1978). Radiation-induced alcohol dehydrogenase mutants in maize following allyl alcohol selection of pollen. *Genet. Res. 31*, 107-129.

Freeling, M., and Schwartz, D. (1973). Genetic relationships between multiple alcohol dehydrogenases of maize. *Biochem. Genet. 8*, 27-36.

Garcia-Bellido, A., and Ripoll, P. (1978). The number of genes in *Drosophila melanogaster*. *Nature 273*, 399-400.

Gleason, M. K., and Metzenberg, R. L. (1974). Regulation of phosphate metabolism in *Neurospora crassa:* Isolation of mutants deficient in the repressible alkaline phosphatase. *Genetics 78*, 645-659.

Goldberg, R. B., Hoshek, G., Kamaley, C., and Timberlake, W. E. (1978). Sequence complexity of nuclear and polysomal RNA in leaves of the tobacco plant. *Cell, 14*, 123-131.

Goldschmidt, R. B. (1951). Chromosomes and genes. *Cold Spring Harb. Symp. Quant. Biol. 16*, 1-11.

Goodman, H. M., Olson, M. V., and Hall, B. D. (1977). Nucleotide sequence of a mutant eucaryotic gene: The yeast tyrosine-inserting ochre suppressor *SUP*4-0. *Proc. Nat. Acad. Sci. (U.S.) 74*, 5453-5457.

Green, M. M. (1967). The genetics of a mutable gene at the white locus of *Drosophila melanogaster, Genetics 56*, 467-482.

Green, M. M. (1969). Controlling element mediated transpositions of the *white* gene in *Drosophila melanogaster*. *Genetics 61*, 429-441.

Green, M. M. (1977a). A case for DNA insertion mutations in *Drosophila melanogaster*. *In* "DNA Insertion Elements, Plasmids and Episomes" (A. Bikhari, J. Shapiro, and A. Adhya, eds.), pp. 434-446. Cold Spring Harbor Press, Cold Spring Harbor, New York.

Green, M. M. (1977b). Genetic instability in *Drosophila melanogaster: De novo* induction of putative insertion mutations. *Proc. Nat. Acad. Sci. (U.S.) 74*, 3490-3493.

Herskowitz, I., Strathern, J. N., Hicks, J. B., and Rine, J. (1977). Mating type interconversion in yeast and its relationship to development in higher eucaryotes. *In* "Eucaryotic Genetic Systems: ICN-UCLA Symposia on Molecular and Cell Biology," Vol. 8 (G. Wilcox, J. Abelson, and C. F. Fox, eds.), pp. 193-202. Academic Press, New York.

Hiraizumi, Y. (1971). Spontaneous recombination in *Drosophila melanogaster* males. *Proc. Nat. Acad. Sci. (U.S.) 68*, 268-270.

Hynes, M. J. (1975). A *cis*-dominant regulatory mutation affecting enzyme induction in the eucaryote *Aspergillus nidulans*. *Nature 253*, 210-211.

Jeffreys, A. J., and Flavell, R. A. (1977a). A physical map of the DNA regions flanking the rabbit β-globin gene. *Cell 12*, 429-440.

Jeffreys, A. J., and Flavell, R. A. (1977b). The rabbit β-globin gene contains a large insert in the coding sequence. *Cell 12*, 1097-1108.

Kabat, D., and Koler, R. D. (1975). The thalassemias: Models for analysis of quantitative gene control. *Adv. in Human Gent. 5*, 157-222.

Knapp, G., Beckman, J. S., Johnson, P. F., Fuhrman, S. A., and Abelson, J. (1978). Transcription and processing of intervening sequences in yeast tRNA genes. *Cell 14*, 221-226.

Lehman, J. F., and Metzenberg, R. L. (1976). Regulation of phosphate metabolism in *Neurospora crassa:* Identification of the structural gene for repressible alkaline phosphatase. *Genetics 84*, 175-182.

Lukaszkiewicz, Z., and Paszewski, A. (1976). Hyper-repressible operator-type mutant on sulphate permease gene of *Aspergillus nidulans*. *Nature 259*, 337-338.

Matthews, K. A., Slatko, B. E., Martin, D. W., and Hiraizumi, Y. (1978). A consideration of the negative correlation between transmission ratio and recombination frequency in a male recombination system of *Drosophila melanogaster*. *Japan. J. Genet. 53*, 13-25.

McClintock, B. (1950. The origin and behavior of mutable loci in maize. *Proc. Nat. Acad. Sci. (U.S.) 36*, 344-355.

McClintock, B. (1951). Chromosome organization and genic expression. *Cold Spring Harb. Symp. Quant. Biol. 16*, 13-48.

McClintock, B. (1967). Genetic systems regulation gene expression during development. *Develop. Biol. Suppl. 1*, 84-112.

Mottinger, J. P. (1970). The effects of x rays on the *bronze* and *shrunken* loci in maize. *Genetics 64*, 259-271.

Mottinger, J. P. (1973). Unstable mutants of *Bronze* induced by pre-meiotic x ray treatment in maize. *Theoret. Appl. Genet. 43*, 190-195.

Muller, J. J. (1955). On the relation between chromosome changes and gene mutations. *Brookhaven Symp. Biol. 8*, 126-145.

Nelson, O. E. (1968). The waxy locus in maize. II. The location of the controlling element alleles. *Genetics 60*, 506-524.

Neuffer, M. G. (1955). Dosage effect of multiple *Dt* loci on mutation at *a1* in maize endosperm. *Science 121*, 399-400.

Peterson, P. A. (1976). Basis for the diversity of states of controlling elements in maize. *Molec. Gen. Genet. 149*, 5-21.

Rand, K. N., and Arst, H. N., Jr. (1977). A mutation in *Aspergillus nidulans* which affects the regulation of nitrite reductase and is tightly linked to its structural gene. *Molec. Gen. Genet. 155*, 67-75.

Rhoades, M. M. (1938). Effect of the *Dt* gene on the mutability of the *a1* allele in maize. *Genetics 23*, 337-397.

Sastry, G. R. K., Cooper, H. B., Jr., and Brink, R. A. (1965). Paramutation and somatic mosaicism in maize. *Genetics 52*, 407-424.

Schwartz, D. (1962). Genetic studies on mutant enzymes in maize. III. Control of gene action in the synthesis of pH 7.5 esterase. *Genetics 47*, 1609-1615.

Schwartz, D. (1966). The genetic control of alcohol dehydrogenase in maize: Gene duplication and repression. *Proc. Nat. Acad. Sci. (U.S.) 56*, 1431-1436.

Schwartz, D. (1969). Alcohol dehydrogenase in maize: Genetic basis for multiple isozymes. *Science 164*, 585-586.

Schwartz, D. (1971). Genetic control of alcohol dehydrogenase --a competition model for regulation of gene action. *Genetics 67*, 411-425.

Stadler, L. J. (1944). The effect of x rays upon dominant mutation in maize. *Proc. Nat. Acad. Cci. (U.S.) 30*, 123-128.

Steeves, T. A., and Sussex, I. M. (1972). "Patterns in Plant Development." Prentice Hall, Englewood Cliffs, New Jersey.

Swank, R. T., Paigen, K., and Ganschow, R. E. (1973). Genetic control of glucuronidase induction in mice. *J. Mol. Biol. 81*, 225-243.

Tartof, K. D. (1971). Increasing the multiplicity of ribosomal RNA genes in *Drosophila melanogaster*. *Science 171*, 294-297.

Thompson, P., and Patel, G. (1972). Compensatory regulation of two closely related loci in *Chironomus tentans*. *Genetics 70*, 275-290.

Thuriax, P., Ramos, F., Piérard, A., Grenson, M., and Wiame, M. M. (1972). Regulation of the carbamylphosphate synthetase belonging to the arginine biosynthetic pathway of *Saccharomyces cerevisiae*. *J. Mol. Biol. 67*, 277-287.

Tilghman, S. M., Tiemeier, D. C., Seidman, T. C., Peterlin, B., Sullivan, M., Maizek, J. V., and Leder, P. (1978). An intervening sequence of DNA identified in the structural portion of a mouse β-globulin gene. *Proc. Nat. Acad. Sci. (U.S.) 75,* 725-727.

Tonegawa, S., Brack, C., Hozumi, N., and Schuller, R. (1977). Cloning of an immunoglobulin variable region gene from mouse embryo. *Proc. Nat. Acad. Sci. (U.S.) 74,* 3518-3522.

Torres, A. M. (1974). Genetics of sunflower alcohol dehydrogenase *Adh2,* nonlinkage to *Adhl* and *Adhl* early alleles. *Biochem. Genet. 12,* 385-392.

Valenzuela, P., Venegas, A., Weinberg, F., Bishop, R., and Rutter, W. J. (1978). Structure of yeast phenylalanine-tRNA genes: An intervening DNA segment within the region coding for the tRNA. *Proc. Nat. Acad. Sci. (U.S.) 75,* 190-194.

Wallace, B., and Kass, T. L. (1974). On the structure of gene control regions. *Genetics 77,* 541-558.

White, R. L., and Hogness, D. S. (1977). R loop mapping of the 18S and 28S sequences in the long and short repeating units of *Drosophila melanogaster* rDNA. *Cell 10,* 177-192.

Wilson, E. B. (1925). "The Cell," Chap. 14. McMillan, New York.

Woodman, J. C. (1978). An organ-specific reciprocal effect demonstrated by *Adhl* alleles. *Maize Genetic Coop. News Lett. 52,* 6-8.

ROLE OF STORED MESSENGER RNA IN LATE EMBRYO DEVELOPMENT AND GERMINATION

L. S. Dure III

Department of Biochemistry
University of Georgia
Athens, Georgia

OUR EARLY FINDINGS IN COTTON SEED DEVELOPMENT

Several years ago we collected evidence that most of the protein synthesis necessary to begin germination in cotton cotyledons is directed by mRNA that is transcribed in embryogenesis (Waters and Dure, 1966; Ihle and Dure, 1969). Further, we were able to show that some of the enzymes translated from this mRNA are unique to germination; that is, they are not enzymes found in embryogenesis. One of these enzymes (a carboxypeptidase C) was purified to homogeneity in order to show by *in vivo* isotope labeling that its appearance in germination is the result of *de novo* synthesis as opposed to zymogen activation, etc. (Ihle and Dure, 1972a, 1972b). Thus, it appears that the mRNA transcribed in embryogenesis but not translated until germination is not simply residual mRNA for omnipresent enzymes that is carried over from embryogenesis into germination, but specific mRNA for proteins necessary for the subsequent developmental event. In this light the conserved mRNA represents a developmentally significant phenomenon; one that can be viewed as part of the preprogrammed equipping of the seed for successful germination.

Subsequently, we were able to implicate the plant growth regulator, abscisic acid (ABA) in the inhibition of the premature translation of this body of mRNA in embryogenesis (Ihle and Dure, 1970, 1972c). This compound, which has been implicated in many inhibitory processes in higher plants, first becomes demonstrable in ovule tissues at the same time that the existence of the "germination mRNA" becomes demonstrable in embryonic cotyledons. Thus, ABA can be viewed as preventing

ISBN 0-12-602050-7

vivipary in these seeds. (Vivipary in higher plants refers to the premature germination of immature seeds while still in the immature fruit. It constitutes a genetic lethal and is found occasionally in almost all angiosperm families). Oddly, it was found that the inhibition of the translation of the germination mRNA by ABA requires continued RNA synthesis (Ihle and Dure, 1970, 1972c). That is, if RNA synthesis is blocked in immature, excised embryos by actinomycin D, the inhibition by ABA of germination enzyme synthesis (as followed by carboxypeptidase activity) is overcome. The immature embryos germinate precociously and carboxypeptidase C activity appears.

Cotton embryos that have not yet reached the point in development at which the germination mRNA becomes demonstrable and ABA appears will also germinate precociously when removed from the ovular tissue. This germination is totally sensitive to actinomycin D as is the appearance of the germination enzymes. The precocious germination of these embryos possibly entails a derepression of the cistrons comprising the germination mRNA. Furthermore, cell division in the cotyledons of these very young embryos is stopped when they are removed from the ovular tissue for germination. Cell division stops *in vivo* in this tissue at the same time that the germination mRNA and ABA appears. Finally, the vascular connection between the incipient seed and the mother plant atrophies naturally at this point. From this it would seem that the vascular flow from the mother plant is what keeps the embryos developing as embryos; that is, maintaining cell division and keeping the cistrons for the germination mRNA repressed. When this vascular flow is stopped normally by the atrophy of the connecting tissue (the finiculus) or mechanically by the excision of very young embryos, embryonic growth (cell division) stops and the germination cistrons are derepressed. *In vivo* the advent of ABA prevents the newly synthesized germination mRNA from being translated during the remaining 20 days that the embryos remain in the cotton boll (maturation phase of embryogenesis).

MORE RECENT OBSERVATIONS

One possible mechanism by which the translation of the germination mRNA is delayed in cotton cotyledons would be a delay in mRNA processing from the initial transcription product to functional mRNA until germination (until the disappearance of ABA). An indication that this may be the case is the effect of 3'd adenosine (3'd Ado, cordycepin) on visible germination and on protein synthesis and the advent of carboxypeptidase activity during germination. In these experiments the effects of 3'd Ado and 3'd cytidine (3'd Cyd) on carboxypeptidase activity were determined in cotyledons that were exposed to these nucleoside analogs for different time periods during the first 36 hr of germination. It was assumed at this time that these analogs would be triphosphorylated *in vivo*, and that the 3'd nucleoside triphosphates would truncate RNA synthesis by becoming incorporated into growing RNA chains as has been shown to take place in a number of tissues (Siev *et al.*, 1969; Abelson and Penman, 1972) and shown to occur in cell-free RNA synthesis by purified RNA polymerases I and II (Blatti *et al.*, 1970; Horowitz *et al.*, 1974).

3'd Cyd, analogous to actinomycin 3'd Cyd, failed to inhibit visible cotyledon germination, gross protein synthesis, or the advent and accumulation of carboxypeptidase activity, whereas 3'd Ado inhibited all of these parameters provided it was supplied to the tissue prior to the thirty-second hour of germination (Walbot *et al.*, 1974). In fact, the difference in the degree of inhibition of carboxypeptidase activity found at the different incubation time periods suggested that the 3'd Ado-sensitive process takes place between the sixth and thirtieth hour of germination.

To establish that the nucleoside analogs through their putative 3'd nucleotide triphosphates do indeed abort the synthesis of all classes of RNA, their effect on the synthesis of rRNA, tRNA, and mRNA-poly(A) during early germination was studied by following isotope incorporation (Harris and Dure, 1974). The use of poly(U)-Sepharose and oligo (dT)-cellulose affinity chromatography made it possible for the first time to quantitatively follow the effect of 3'd Cyd on mRNA-poly(A) synthesis. Since 3'd Ado presumably leads to the truncation of polyadenylation, its effect on mRNA synthesis could not be followed, since the sequestration and concentration on the affinity columns of mRNA synthesized during the labeling period cannot occur in the absence of the poly(A) region of the chain.

In the case of rRNA (including 5 S RNA) and tRNA, both nucleoside analogs caused essentially complete inhibition of isotope incorporation in germinating cotyledons. However, to our surprise, 3'd Cyd had no effect on the synthesis of mRNA-poly(A). By analogy, 3'd ATP derived from 3'd Ado is also ineffective in truncating mRNA synthesis, since it seems unlikely that the polymerase can sense the lack of a 3' OH group on one nucleotide triphosphate but not on another. However, this presumed synthesis cannot be demonstrated by these methods if the poly(A) moiety is missing, as mentioned above. These data indicate a difference in substrate recognition between RNA polymerases I and II. It has been shown in HeLa cells that the synthesis of HnRNA, the presumed initial transcription product of RNA polymerase II, is also not affected by the nucleoside analogs (Siev *et al.*, 1969; Abelson and Penman, 1972). Yet 3'd ATP has been shown to cause chain termination in cell-free transcription by purified RNA polymerase II. This curious inconsistency has not been resolved to our knowledge.

The net effect of the latter experiments showing that 3'd Cyd does not cause an inhibition of mRNA synthesis was that we were once again in the position of juxtaposing the effects of actinomycin D and 3'd Ado on visible germination, protein synthesis, and carboxypeptidase activity. These results taken at face value indicate that although the synthesis of some of the germination proteins in cotyledons during the first 3 days of germination may not require mRNA synthesis, it does require polyadenylation.

The ability to sequester mRNA-poly(A) on the affinity columns made it feasible to test directly the possibility that poly(A) is added onto preformed RNA during the early stages of germination. This idea received some additional reinforcement when we measured the effect of actinomycin on mRNA-poly(A) synthesis using the affinity columns. Dosages of actinomycin that inhibit essentially all rRNA and tRNA synthesis were found to stop about 70% of isotope incorporation into mRNA but only 30% of isotope incorporation into poly(A). If only newly synthesized mRNA is polyadenylated during early germination, isotope (^{32}P) incorporation into poly(A) should be inhibited to the same extent as is its incorporation into mRNA; that is, only the mRNA that is synthesized during the labeling period should get polyadenylated. In actuality much more gets polyadenylated.

In view of the above suggestive evidence that preformed mRNA becomes polyadenylated during early germination, we designed experiments that allowed us to measure the incorporation of ^{32}P and ^{3}H-adenosine into mRNA-poly(A) and into the mRNA and poly(A) portions of these molecules individually. The rationale of these experiments was that if the polyadenylation of some or all of the preformed mRNA takes place during the labeling period, the mRNA portion of the mRNA-poly(A) fraction will be underlabeled relative to the poly(A) portion. That is, any preformed mRNA that is polyadenylated during the labeling period will contain isotopes only in the poly(A) component. When the apparent mass average chain length of the mRNA portion is then calculated from its isotope content relative to the isotope content of the poly(A) portion and the value obtained is compared with the true average chain length obtained from gel electrophoresis under fully denaturing conditions (in 99% formamide), the difference between these values should be an indication of the ratio of preformed to newly synthesized mNRA that is polyadenylated during the labeling period.

If these data are obtained from cotyledons germinated in the presence of actinomycin D, any apparent underlabeling of the mRNA portion of the mRNA-poly(A) fraction should be accentuated, since 70% of the synthesis of new mRNA will have been inhibited, which, in turn, should confine most of the isotope incorporation to the poly(A) portion of preformed mRNA.

Conversely, if these same measurements of the distribution of radioactivity between mRNA and poly(A) are made at a time in germination after the putative stored mRNA has presumably been processed into functional polysomal mRNA, the distribution of radioactivity should show no underlabeling of mRNA, and the calculated mRNA mass average chain length should be consonant with that found by gel electrophoresis under fully denaturing conditions.

The results of these experiments upheld these predictions (Harris and Dure, 1976; Dure and Harris, 1977; Harris and Dure, 1978). The calculated mass average chain length of mRNA was absurdly low for mRNA labeled during the first 24 hr of germination, even lower for mRNA from actinomycin D-treated cotyledons, but close to the real lengths (*ca.* 1800 nucleotides) in the mRNA labeled later in germination. These data allowed us further to calculate that from 55 to 65% of the mRNA polyadenylated during the first day of germination existed in the dry seed, i.e., is stored mRNA presumably encoding the germination enzyme sequences. This value was further substantiated by experiments in which the increase in mRNA concentration occurring in actinomycin D-treated cotyledons was measured optically and by hybridization with ^{14}C-poly(U). It should be noted that these measurements measure the mass of mRNA polyadenylated and not its sequence complexity. Thus the 55-65% of mRNA that is apparently stored mRNA may represent only a small percentage of the genetic complexity of the total mRNA polyadenylated. That is, the number of different proteins resulting from the stored mRNA may be only a small number, but would belong to the abundant class. The difference in cellular concentration of different mRNAs has been found in many instances to be enormous (Axel *et al.*, 1976).

The results of these experiments further substantiated the idea that some or all of the stored cotyledon mRNA is not processed until germination and that this failure to process the germination mRNA in enbryogenesis is an aspect of the mechanism by which its premature translation in embryogenesis is prevented. However, such experiments are not definitive. Factors that are as yet unperceived could be responsible for the observed distribution of radioactivity between mRNA and poly(A) during the various labeling periods. The definitive experiment would be to prove that mRNA sequences that are polyadenylated during germination exist in the dry seed and in immature embryos in a nonfunctional form. This requires that sequence identity be established by DNA:RNA hybridization and DNA:cDNA reassociation experiments.

STORED VERSUS RESIDUAL mRNA

The conceptual distinctions between stored mRNA and the residual mRNA of the dry seed embryo should be kept in mind in devising and interpreting experiments dealing with seed germination. Both types of mRNA can be considered "preformed" mRNA in that they are initially transcribed in embryogenesis, survive desiccation, and participate in protein synthesis in early germination. The critical distinction between these mRNA subsets is that residual mRNA is a remnant of embryonic growth and codes for proteins that characterize embryonic growth. Stored mRNA, on the other hand, is considered to be an aspect of the embryo's preparation for germination. That is, it does not participate in embryonic growth and codes for proteins that characterize germination growth. There is growing evidence for the existence of residual mRNA in mature seeds (Dure, 1977), whereas stored mRNA is a theoretical entity we have been forced to postulate to explain our experimental observations. The following outlines list characteristics of these two mRNA subsets and present the evidence for their existence in cotton cotyledons.

Postulated Properties of Residual mRNA
1. Transcribed in embryogenesis
2. Translated in embryogenesis and first hours of germination
3. Destroyed in early germination

Evidence For Residual mRNA
1. Immediate resumption of protein synthesis upon germination in actinomycin D
2. Storage protein synthesized in first hours of germination
3. Existence of mRNA-poly(A) in dry seeds

Postulated Properties of Stored mRNA
1. Transcribed in embryogenesis
2. Translation prevented by ABA
3. Resultant proteins unique to germination

Evidence for Stored mRNA
1. Failure of actinomycin D to inhibit protein synthesis, 0-48 hr of germination
2. Carboxypeptidase C activity insensitive to actinomycin D, sensitive to 3'd Ado during

germination, sensitive to ABA in precocious germination
3. Nontheoretical distribution of isotopes between mRNA and poly(A) during first 20 hr of germination
4. Increase in mRNA-poly(A) in actinomycin D during first 20 hr of germination

A further postulate pertaining to these subsets is that residual mRNA should be present in all seed types, whereas the putative stored mRNA may be restricted to dicots that carry embryonic development to completion; i.e., that totally consume the endosperm/nucellar layers and store all their nutrient reserves in the embryo itself. A more extensive development of these ideas and their experimental underpinning can be found in Dure (1977).

POSSIBLE mRNA SUBSETS IN THE EMBRYOGENESIS AND GERMINATION OF COTTON COTYLEDONS

Since we were forced (by indirect evidence only) to conjecture that mature cotton cotyledons contain two distinct coexisting subsets of mRNA, it became apparent that even more subsets could be predicted when cotton seed development is examined from a larger perspective, as in Table I.

These putative subsets, again resting solely on indirect evidence, could conceivably differ one from another not only in the time and/or rate of their transcription, but also in their processing, translation, or degradation. This assumes that cotyledon cells recognize mRNA (or pre-mRNA) members of the different subsets and treat them accordingly. In turn this implies that in some fashion members of each subset differ chemically from members of other subsets.

CURRENT RESEARCH

Because we have tentatively postulated the mRNA subsets and implied that their unique features play a role in the regulation of development, we have set out to measure the population of mRNA in the various subsets directly and indirectly by describing the protein domains specified by each subset. In this we are attempting to separate and identify the proteins that are extant during the various stages of embryogenesis and germination and to identify proteins that appear to represent products of the several mRNA subsets. Our cataloging process is presented in Table II.

TABLE I. Possible mRNA subsets

Embryogenesis (after cessation of cell division)	Representative protein
1. Functional during ABA arrestation	storage protein
2. Nonfunctional in ABA arrestation	carboxypeptidase C
3. Disappears upon ABA arrestation	?
4. Constitutive (independent of developmental events)	?
Germination	
1. Residual	storage protein
2. Stored	carboxypeptidase C
3. Newly synthesized, unique to germination	asparagine synthetase[a]
4. Newly synthesized, constitutive	?

[a]*The listing of asparagine synthetase as a representative product of newly synthesized mRNA is from Dilworth and Dure (1978).*

TABLE II. Determination of RNA Sequences Existing in Late Embryogenesis, in Mature Seeds, and in Germinating Cotyledons (± Act D)

1. Display of existing proteins	by gel electrophoresis
2. Display of proteins synthesized in vivo	by fluorography of gels
3. Display of proteins synthesized in vitro from total RNA and polysomal RNA	by fluorography of gels
4. Determination of RNA complexity and complexity overlap	by RNA:cDNA hybridization

Although the rather arbitrary listing of possible developmental subsets is not meant to be taken too seriously, it does provide a framework upon which to devise further experimentation. If such subsets do exist then defining their chemical uniqueness would be a giant step in coming to grips with the regulation of development. It is against this background that we are carrying out the current lines of investigation in our laboratory.

The first of the projects simply involves extracting the total protein from cotyledons at different developmental stages with solutions high in SDS and mercaptoethanol and separating them by discontinuous SDS gel electrophoresis (Laemmli, 1970) or by electrophoresis in two dimensions (O'Farrell, 1975). Figure 1 is an example of such a separation. In this case total protein was extracted from cotyledons at various intervals of embryogenesis and germination and displayed by electrophoresis in one dimension. The embryogenesis numbers refer to the weight in mgs of the cotyledons when extracted, and the germination numbers refer to days of germination. "L" denotes cotyledonary protein from seedlings that were greened, whereas "D" refers to that from dark grown seedlings.

This technique, of course, only displays the abundant class of proteins, and the thousands of enzymes whose cellular concentration is likely to be low are not observed. Because many of the abundant embryogenic proteins are likely to be stable storage proteins, such a display does not necessarily reflect the mRNA population functioning at any of the developmental points. Nevertheless, such displays do provide information that may be useful on a gross level in identifying mRNAs of some of the putative subsets. For example, it is apparent that the smaller of the two dominant storage proteins (see dry seed profile) appears first and accumulates faster than does the larger storage protein. It is also degraded much more rapidly during germination. Furthermore, two proteins that are abundant in embryogenesis (arrows) vanish during the last stage of embryogenesis and are undetectable in dry seed cotyledons. It is unfortunate that few readily observable, abundant proteins accumulate during germination. The abundant protein shown by the arrow has proven to be a degradation product of the smaller storage protein. The RUDP subunits (*ca.* 55,000 and 12,000 mol. wt.) can be distinguished in the profiles of protein from 4, 5, and 6 day germinated cotyledons.

When embryonic cotyledons are incubated in radioactive amino acids for several hours, the proteins extracted and electrophoresed, and the stained electrophorogram fluorographed to detect radioactive bands, a better indication of the mRNA population can be obtained. Data we have obtained in this area (Project 2) have presented some surprises, as shown in Fig. 2. Here protein extracted from 100 mg cotyledons exposed to iso-

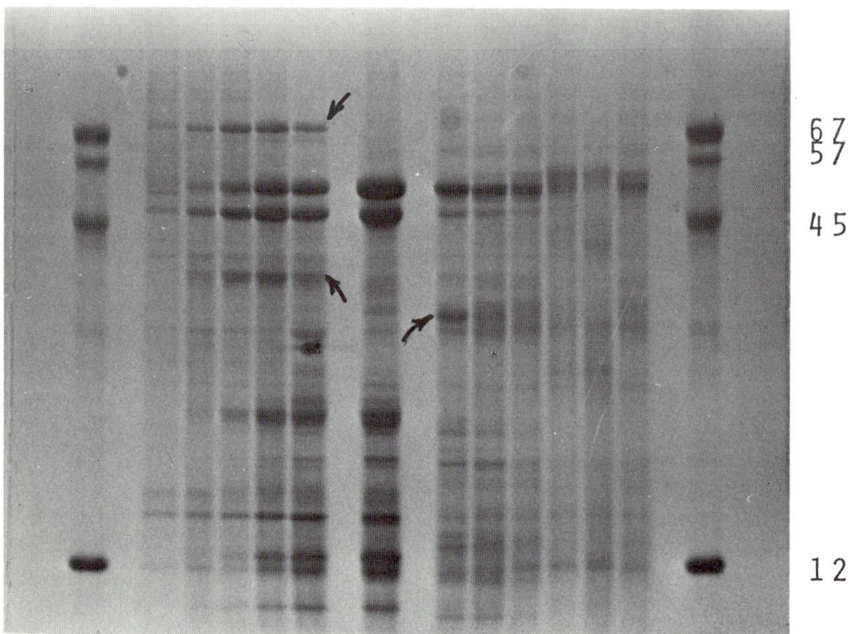

FIGURE 1. One dimensional gel electrophoretic separation of total protein extracted from cotton cotyledons at different developmental stages. Numbers at the top refer to the developmental stages as given in the text. Numbers on the right side refer to the molecular weight in thousands of daltons of proteins run as standards in the extreme right and left hand wells. The electrophoretic system was the discontinuous polyacrylamide SDS system of Laemmli (1970) using an acrylamide gradient of 10-17%; gels were stained with Coomassie R 250.

tope for 6 hr has been electrophoresed in one dimension and the gel fluorographed. (These embryos will germinate precociously and the synthesis pattern will resemble that of germinated mature cotyledons after several days. Thus in this case the labeling period was relatively short in order to show embryo-

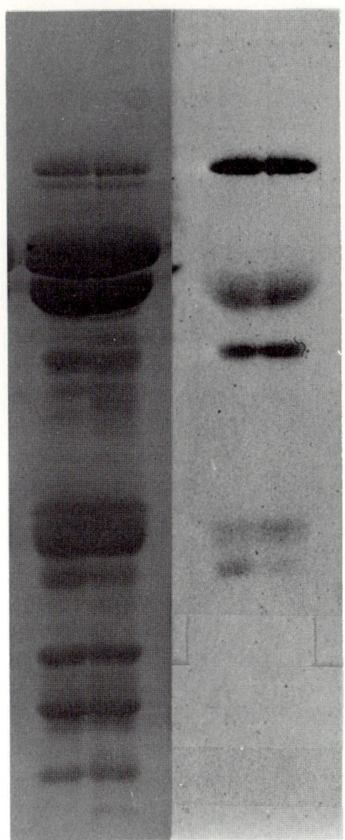

FIGURE 2. In vivo protein synthesis by cotton cotyledons during last phase of embryogenesis. Embryos of 100 mg wet weight were dissected and exposed to ^{14}C-amino acids for 6 hr after which total protein was extracted and separated in one dimension as in Fig. 1. Left-hand wells are of stained gel; right-hand profile is the fluorograph of the left-hand wells.

genic synthesis). The stained gel (left two wells of the gel) shows the position of the abundant proteins that have accumulated. The two dominant storage proteins are quite obvious. The gel is overloaded in order to contain enough radioactivity. The right-hand wells, fluorographs of the left-hand wells, show that the 70,000 mol. wt. protein and the smaller of the two dominant storage proteins are being synthesized in these cotyledons at this stage of embryogenesis. Yet the 70,000 mol. wt. protein is disappearing during this period and the larger storage protein is accumulating faster than the smaller in order to give the relative abundancy seen in mature seed cotyledons. Obviously these data were not anticipated and suggest that there may be a number of events taking place between the synthesis of initial nascent protein and the ultimate accumulation of "finished" protein. A number of possible explanations of the data come to mind, such as a precursor-product relationship between the smaller and larger storage protein. The fact that the larger storage protein is highly glycosylated makes this idea rather attractive. Should this notion be correct, it should be possible to "chase" radioactivity into the larger protein by extending the incubation period. However, we have not yet pieced together the sequence of events that reconcile the stain versus radioactivity data in embryogenesis. This must be worked out before mRNAs representative of this period can be identified.

Data from *in vivo* translation studies have reinforced the earlier observations that in the first several days of germination actinomycin D blocks the synthesis of new mRNA (by 70%) but has no observable effect on the total incorporation of radioactive amino acids into total protein. Figure 3 demonstrates this. In this gel, well 1 was loaded with a total protein extract from cotyledons germinated 24 hr in radioactive amino acids and well 2 contains the equivalent from cotyledons germinated in actinomycin D. Most of the discernible labeled bands are found in both wells, indicating that these proteins probably originate from residual and/or stored mRNA. The two areas indicated by arrows contain radioactive bands only in the protein preparation from untreated cotyledons. Our present interpretation is that these labeled proteins originate from mRNA newly synthesized during the first day of germination.

The data we have obtained to date by electrophoretic separation of the radioactive proteins that are translated from purified mRNA *in vitro* (Project 3) have been somewhat difficult to equate to similar data obtained from *in vivo* labeling. This again suggests there is a good deal of protein processing that takes place *in vivo* but not in the cell-free translation system. We must sort out much of this before the stained gels, the *in vivo* labeled profiles, and the *in vitro* labeled profiles become interpretable in a meaningful way. Only then can assignments of mRNA to the putative subsets be made and their putative unique features be approached experimentally.

Project 4 attempts to characterize the various mRNA populations during development through their hybridization kinetics with cotton DNA and with cDNA synthesized from the mRNA populations themselves. These experiments cannot result in the assignment of specific mRNA and their resultant protein products to the putative subsets, but they can give an idea of the number of gene transcripts active in cotyledons at different points in development and the degree to which the total mRNA population changes during development. In fact, this last is

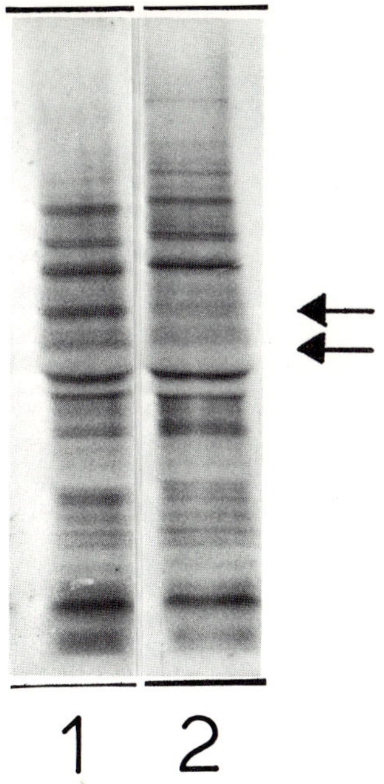

FIGURE 3. In vivo protein synthesis by cotton cotyledons during the first 24 hr of germination. Embryos were dissected, germinated for 24 hr and exposed to ^{14}C-amino acids between the 10th and 24th hr. Total protein was extracted from cotyledons and separated by gel electrophoresis as given in Fig. 1. The gel was autoradiographed to give the distribution of radioactivity. Well 1 contained protein from normally germinating cotyledons, whereas well 2 contained an identical aliquot of protein from cotyledons germinated in actinomycin D at 20 μg/ml.

the only project of the four that measures the total genetic complexity of the mRNA populations. Hybridization of mRNA against cDNA synthesized from it should allow for measurements of the number of members of the abundant class of mRNAs (and thus of proteins). It is anticipated that the bulk of the mRNA complexity (which probably codes for housekeeping or maintenance enzymes) will be common to all stages of development.

Our data to date from this project is still beset with unresolved problems which we hope will prove solvable in the near future.

REFERENCES

Abelson, H. T., and Penman, S. (1972). Messenger RNA formation: resistance to inhibition by 3'-deoxycytidine. *Biochim. Biophys. Acta 277,* 129-133.

Axel, R., Feigelson, P., and Schutz, G. (1976). Analysis of the complexity and diversity of mRNA from chicken liver and oviduct. *Cell 7,* 247-254.

Blatti, S. P., Ingles, C. J., Lindall, T. J., Morris, P. W., Weaver, R. F., Weinberg, F., and Rutter, W. J. (1970). Structure and regulatory properties of eucaryotic RNA polymerase. *Cold Spring Harbor Symp. Quant. Biol. 35,* 649-658.

Dilworth, M. F., and Dure, L. S. III (1978). Developmental biochemistry of cotton seed embryogenesis and germination X. Nitrogen flow from arginine to asparagine in germination. *Plant Physiol. 61,* 698-702.

Dure, L. S. III (1977). Stored messenger ribonucleic acid and seed germination. *In* "The Physiology and Biochemistry of Seed Dormancy and Germination" (A. A. Khan, ed.), pp. 335-345. Elsevier/North Holland Biomedical Press, Amsterdam.

Dure, L. S. III, and Harris, B. (1977). Regulation of translation of stored mRNA of dicot seeds. *In* "Nucleic Acids and Protein Synthesis in Plants" (L. Bogorad and J. H. Weil, eds.), pp. 279-291. Plenum Press, London.

Harris, B., and Dure, L. S. III (1974). Differential effects of 3'-deoxy nucleosides on RNA synthesis in cotton cotyledons. *Biochem. 13,* 5463-5467.

Harris, B., and Dure, L. S. III (1976). Polyadenylation of stored mRNA in cotton seed germination. *In* "Progress in Nucleic Acid Research and Molecular Biology" Vol. 19, (W. Cohn, and R. Volkin, eds.), pp. 113-118. Academic Press, New York.

Harris, B., and Dure, L. S. III (1978). Developmental regulation in cotton seed germination: polyadenylation of stored messenger RNA. *Biochem. 17,* 3250-3256.

Horowitz, B., Goldfinger, B., and Marmur, J. (1974). Effects of cordycepin (COR) and its triphosphate (COR-P) on RNA synthesis in yeast and yeast extracts. *Fed. Amer. Soc. Exp. Biol. 33,* 1418.

Ihle, J. N., and Dure, L. S. III (1969). Synthesis of a protease in germinating cotton cotyledons catalyzed by mRNA synthesized during embryogenesis. *Biochem. Biophys. Res. Comm. 36,* 705-710.

Ihle, J. N., and Dure, L. S. III (1970). Hormonal regulation of translation inhibition requiring RNA synthesis. *Biochem. Biophys. Res. Comm. 38,* 995-1001.

Ihle, J. N., and Dure, L. S. III (1972a). The developmental biochemistry of cottonseed embryogenesis and germination. I. Purification and properties of a carboxypeptidase from germinating cotyledons. *J. Biol. Chem. 247,* 5034-5040.

Ihle, J. N., and Dure, L. S. III (1972b). The developmental biochemistry of cottonseed embryogenesis and germination. II. Catalytic properties of the cotton carboxypeptidase. *J. Biol. Chem. 247,* 5041-5047.

Ihle, J. N., and Dure, L. S. III (1972c). The developmental biochemistry of cottonseed embryogenesis and germination. III. Regulation of the biosynthesis of enzymes utilized in germination. *J. Biol. Chem. 247,* 5048-5055.

Laemmli, U. K. (1970). Cleavage of structural proteins during the assembly of the head of bacteriophage T4. *Nature 227,* 680-685.

O'Farrell, P. H. (1975). High resolution two-dimensional electrophoresis of proteins. *J. Biol. Chem. 250,* 4007-4021.

Siev, M., Weinberg, R., and Penman, S. (1969). The selective interruption of nucleolar RNA synthesis in HeLa cells by cordycepin. *J. Cell Biol. 41,* 510-520.

Walbot, V., Capdevila, A., and Dure, L. S. III (1974). Action of 3'd adenosine (cordycepin) and 3'd cytidine on the translation of the stored mRNA of cotton cotyledons. *Biochem. Biophys. Res. Comm. 60,* 103-110.

Waters, L. C., and Dure, L. S. III (1966). Ribonucleic acid synthesis of germinating cotton seeds. *J. Mol. Biol. 19,* 1-27.

HORMONAL CONTROL OF STORAGE PROTEIN SYNTHESIS IN *PHASEOLUS VULGARIS*[1]

I. M. Sussex
R. M. K. Dale

Department of Biology
Yale University
New Haven, Connecticut

The angiosperm seed is an excellent experimental system for the study of developmental regulation. The seed passes through a succession of well-defined morphological stages that can be used to identify the timing of specific developmental and biochemical events. Its development from ovule to mature seed appears to be largely independent of informational cues from the maternal plant and from the external environment, and is therefore primarily under genetic control. In cultivated species, especially, there exists a large number of varieties, cultivars, and mutants in which the expression of seed development varies according to such factors as the timing of developmental events, the size of the mature seed, or its biochemical composition, thus providing opportunity for study of the genetic components of regulation. Finally, the immature embryo of many species can be removed from the seed and grown in sterile culture where the role of specific substances in its development can be examined.

In the past decade there has been an upsurge of interest in studying seed storage proteins and the regulation of their synthesis. From these studies has emerged the recognition that the amount of any particular protein accumulated per unit time in the seed differs at different stages of embryogeny, and that the relative amounts of the several proteins accumulated vary during seed development within a species, and may be different in different genotypes.

[1]*Research supported by NIH grant GM 24775 and a NIH Postdoctoral Fellowship awarded to R. M. K. Dale.*

ISBN 0-12-602050-7

The amino acid composition of a storage protein has a marked effect on its dietary value. Because plant proteins are typically deficient in one or more of the 10 amino acids that humans and monogastric animals must derive from their diet (because they cannot synthesize them), alteration of seed protein content to one in which there is a better balance of the essential amino acids has great practical importance. As knowledge of the regulatory mechanisms for seed protein synthesis becomes more complete, ways may be found to modify the action of certain of these mechanisms through genetic manipulation or agricultural practice, thereby causing an improvement of protein content in relation to particular needs.

This paper reviews studies that we have carried out to examine the hormonal regulation of the accumulation and synthesis of storage protein in the developing embryo of the bean *Phaseolus vulgaris* cv. Taylor's Horticultural.

DEVELOPMENTAL TIMETABLE

Plants were grown at the Marsh Botanical Garden in New Haven during the months of June to August, or in growth rooms at 25°C, 65 ± 5% relative humidity, 16/8 hr light/dark cycles, with 4 plants in vermiculite per 6 in. plastic pot on watering tables onto which was pumped a nutrient solution of Hyponex (Hyponex Co.) four times per day, each for 30 min. Under these conditions flowering began 28 days after planting, and seeds matured 36 days after flower anthesis. Length and color characteristics of the pod, seed, and embryo formed the basis of a developmental timetable (Walbot *et al.*, 1972) that was used for collections of embryos of uniform developmental age for biochemical and sterile culture experiments.

NORMAL DEVELOPMENT OF THE EMBRYO

Day 20 marks the transition between two contrasting growth phases of the embryo (Fig. 1, and Walbot *et al.*, 1972). Prior to that time water content is high, and dry weight and protein content are low. After day 20 water content decreases progressively, and dry weight and protein content both increase rapidly. Total protein content per unit weight is low before day 20 and increases rapidly after that time (Table I). Sun *et al.* (1978) have shown that in cv. Tendergreen more than 60% of the protein accumulated during this late stage of embryogeny is a single storage protein. The question that we shall examine here is, What controls the change from a low to a high rate of storage protein accumulation in the embryo?

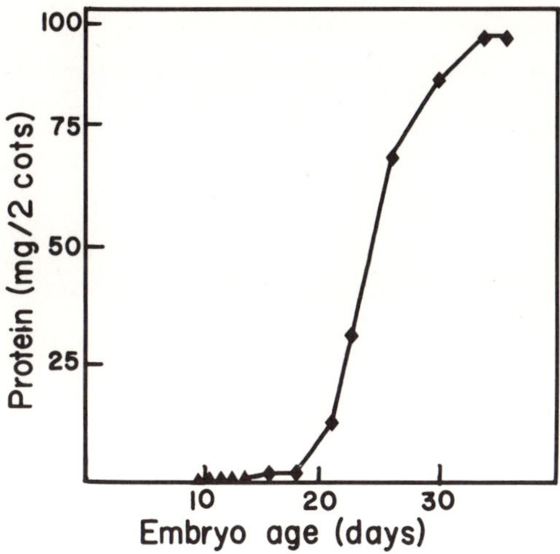

FIGURE 1. Total protein content of cotyledons during embryogeny. Protein was determined from tissue homogenized in 0.1N NaOH by a modified Lowry procedure (Hartree, 1972).

TABLE I. Relative Protein Content During Embryogeny

Embryo age (days)	*Protein/wet wt. (μg/mg)*	*Embryo age (days)*	*Protein/wet wt. (μg/mg)*
10	*17.0*	*21*	*45.3*
11	*14.8*	*22*	*87.3*
12	*18.6*	*26*	*133.0*
13	*21.9*	*30*	*148.0*
14	*25.3*	*34*	*191.0*
15	*35.8*	*36*	*180.0*
16	*34.5*		
18	*37.0*		

SYNTHESIS AND ACCUMULATION OF STORAGE PROTEIN

The major storage protein in the seed of *Phaseolus vulgaris* is Glycoprotein II (Gl II) (Pusztai and Watt, 1970). What is probably the same protein has been isolated from other bean cultivars by different procedures by Racusen and Foote (1971), by McLeester *et al.* (1973), who named the protein Gl, and by Bollini and Chrispeels (1978), who referred to the protein as vicilin. We isolated Gl II from mature seeds of *P. vulgaris* cv. Taylor's Horticultural and purified it to at least 99% purity (Dale, unpublished data). The purified protein was injected into 8 lb female New Zealand white rabbits (Axelsen *et al.*, 1973). From the immune serum a monospecific antibody against Gl II was prepared (Axelsen *et al.*, 1973). This was used to examine the timing and extent of Gl II accumulation during embryogeny by rocket immunoelectrophoresis (Axelsen *et al.*, 1973).

The youngest embryos included in this study were 9 days old and contained a low level of Gl II (Fig. 2). In embryos younger than 20 days Gl II content was low, but in embryos older than day 20 there was a large and continuing increase in Gl II con-

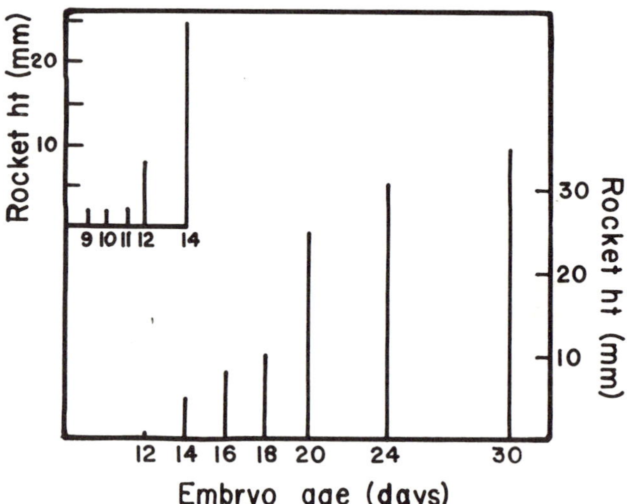

FIGURE 2. Glycoprotein II content of embryos during seed development. Embryos of uniform developmental age were homogenized in a volume of Svendsen buffer (BioRad) equal to embryo wet weight, and 4 μl of centrifuged homogenate supernatant was run on rocket immunoelectrophoresis plates. Inset at upper left shows glycoprotein II content of 9-14 day embryos. This rocket procedure was run at increased sensitivity.

tent (Fig. 2). To examine the basis of this change in the accumulation pattern of Gl II, embryos were removed aseptically from the seed and grown in sterile culture either in 25 ml conical Pyrex flasks in 5 ml of liquid medium, or in 5 cm Falcon plastic petri dishes on 8 ml of agar-solidified medium, at 25°C, and 16/8 hr light/dark cycles. Flask cultures were agitated on a shaker with a 2.5 cm reciprocal motion of 30 cycles per min. The basal culture medium contained the major and minor salts and organic nitrogen components of Mante and Boll (1975), 3% sucrose, and 0.8% agar (as required). The pH was adjusted to 5.5 with NaOH and the medium was autoclaved at 121°C for 20 min. Hormone additions were made to the cooling medium by Millipore sterile filtration.

Nine-day embryos cultured on the basal medium increased in wet weight about 3 times in a 6 day culture period, and Gl II content, as measured by rocket immunoelectrophoresis, approximately doubled (Fig. 3). Addition of the hormones indole acetic acid (IAA), kinetin (K), or gibberellic acid (GA) at four different concentrations had little effect on growth or Gl II content of the embryos (Fig. 3). When the culture medium contained the hormone abscisic acid (ABA), embryo growth was inhibited at all concentrations tested, and embryos cultured on media containing 10^{-6} *M* ABA and higher concentrations had significantly higher Gl II content than those on the basal medium (Fig. 4). The rocket immunoelectrophoresis procedure as used in this study measures concentration of Gl II in the embryo. Because ABA also inhibited embryo growth it was necessary to determine whether embryos cultured on high ABA media contained greater *amounts* of Gl II, whose accumulation was therefore stimulated by ABA, and not simply a greater *concentration* resulting from the inhibition of embryo growth. The data in Tables II and III indicate that IAA, K, and GA do not significantly affect the amount of Gl II in cultured embryos, but that embryos cultured on media containing 10^{-6} to 10^{-4} *M* ABA contain significantly greater amounts of Gl II than do embryos cultured on the basal medium, and that at 10^{-5} *M* ABA there is an approximately 3-fold increase in Gl II amount. Thus ABA at some concentrations stimulates Gl II accumulation.

The next question investigated was whether the increased accumulation of Gl II in embryos cultured on 10^{-5} *M* ABA was the result of increased rate of synthesis or whether it could be accounted for in some other way, such as a reduced rate of degradation of the protein under these conditions of culture. Nine-day embryos were grown in sterile culture for 3 days on basal medium, or with 10^{-5} *M* ABA. The embryos were then incubated for 12 hr in ^{14}C-mixed amino acids (New England Nuclear) (50 µCi/ml of culture medium), washed in fresh culture medium, and returned to fresh culture medium of the same composition. At successive times embryos were processed for analysis of Gl II by scintilla-

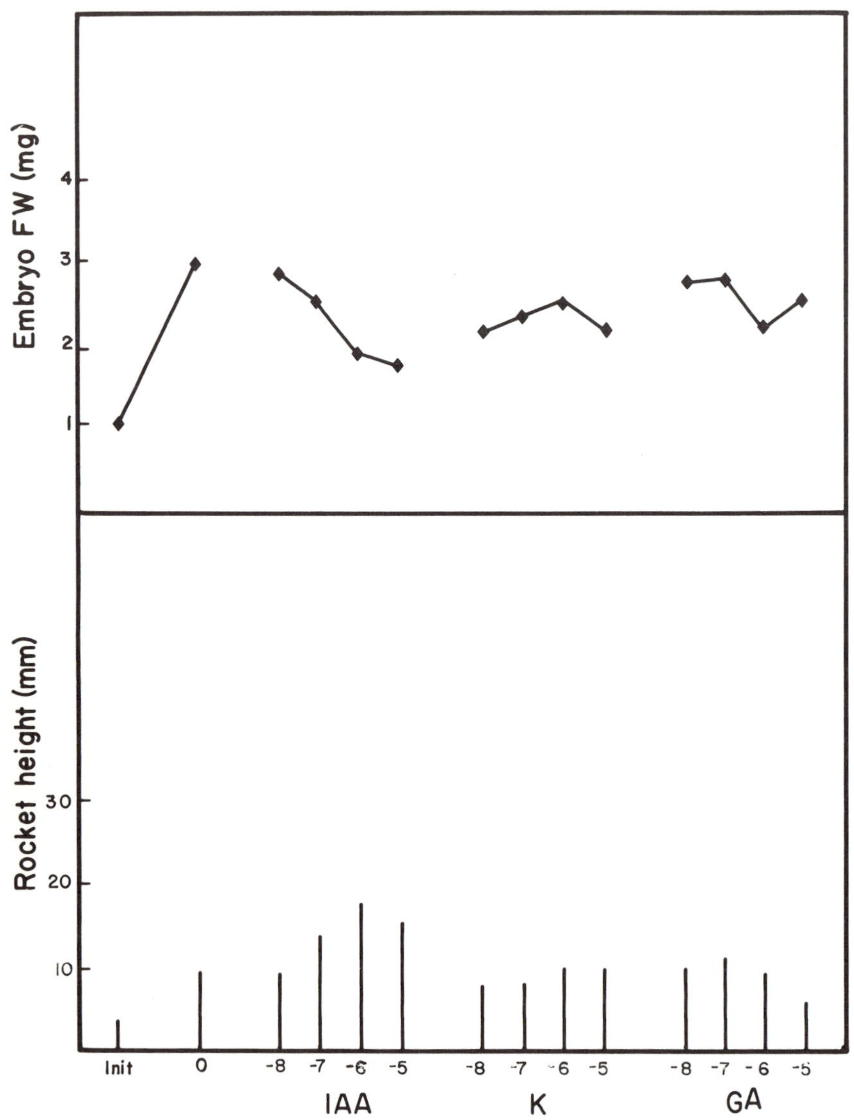

FIGURE 3. Growth (upper panel) and glycoprotein II content (lower panel) of 9 day embryos removed from the seed (init) and those cultured for 6 days on basal medium (0), or on basal medium supplemented with 10^{-8}*-*10^{-5} M *IAA, K, or GA.*

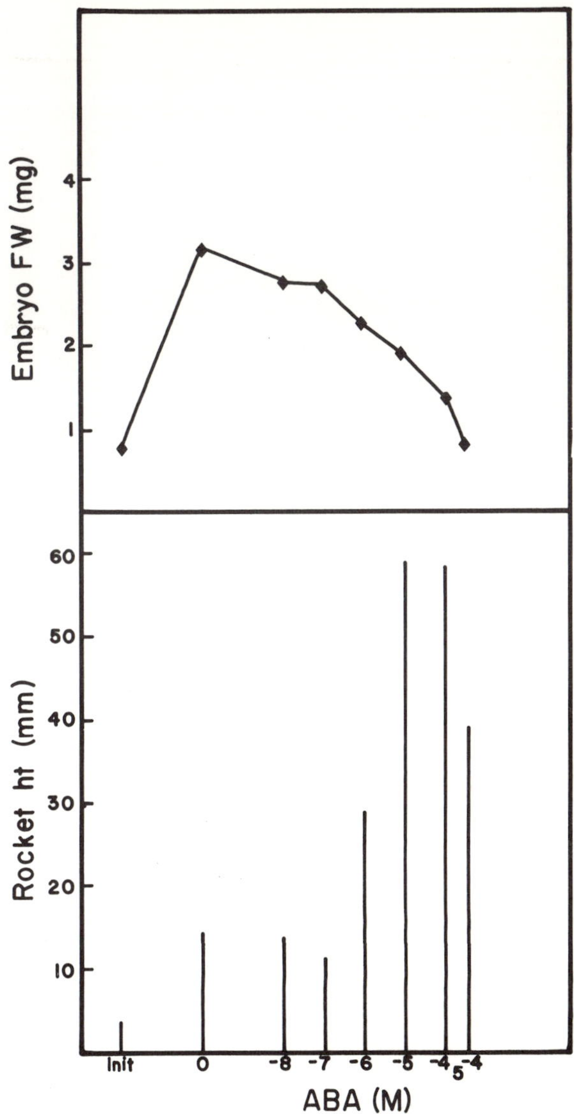

FIGURE 4. Growth (upper panel) and glycoprotein II content (lower panel) of 9 day embryos removed from the seed (init) and those cultured for 6 days on basal medium (0), or on basal medium supplemented with 10^{-8}*-*5×10^{-4} *M ABA.*

TABLE II. Glycoprotein II in Cultured Embryos

Medium addition	*Gl II equivalents per embryo (μg)*
Initial	*3.5*
basal medium	*28.2*
IAA 10^{-8} *M*	*27.2*
IAA 10^{-7} *M*	*34.5*
IAA 10^{-6} *M*	*34.3*
IAA 10^{-5} *M*	*31.3*
K 10^{-8} *M*	*17.4*
K 10^{-7} *M*	*18.7*
K 10^{-6} *M*	*25.6*
K 10^{-5} *M*	*21.7*
GA 10^{-8} *M*	*27.4*
GA 10^{-7} *M*	*30.7*
GA 10^{-6} *M*	*21.0*
GA 10^{-5} *M*	*16.7*

tion counting of the Gl II "rockets" which were cut out of the electrophoresis plates and counted in Aquasol. The data in Table IV show that at 7 and 14 days after the embryos had been radioactively labeled, Gl II from embryos cultured on ABA contains approximately twice as much radioactivity as Gl II from ungerminated embryos in basal medium, and that between day 7 and day 14 there was not a significant decrease in Gl II radioactivity in embryos on either culture medium. By day 14 approximately 50% of the embryos cultured on the basal medium had

TABLE III. Glycoprotein II in Cultured Embryos

Medium addition	*Gl II equivalents per embryo (μg)*
Initial	*3.2*
basal medium	*47.0*
ABA 10^{-8} *M*	*39.0*
ABA 10^{-7} *M*	*31.0*
ABA 10^{-6} *M*	*66.0*
ABA 10^{-5} *M*	*117.0*
ABA 10^{-4} *M*	*88.0*
ABA 5×10^{-4} *M*	*31.0*

TABLE IV. Stability of Glycoprotein II in Cultured Embryos[a]

	Basal medium (cpm)	*+ABA* 10^{-5} M *(cpm)*
day 7	*10,990*	*23,349*
day 14	*11,897*	*21,470*
germinated	*5784*	-

[a]*Embryos labeled for 12 hr with 50 μCi/ml* ^{14}C*-mixed amino acids.*

germinated, as indicated by elongation of the primary root, development of root hairs, and the start of primary leaf expansion. Gl II from germinated embryos contained less radioactivity than was present in Gl II from ungerminated embryos on the same medium, indicating that during germination Gl II is degraded. This is consistent with its known metabolism during seed germination (Bollini and Chrispeels, 1978). From this experiment it appears that Gl II is not degraded to a significant amount in cultured embryos prior to their germination and that the stimulation caused by ABA reflects stimulation of the rate of synthesis.

The next question examined was whether ABA is required continuously for stimulation of Gl II synthesis, or whether brief exposure to ABA results in a persistent increase in Gl II synthesis. Nine day embryos were cultured for 3 days either on basal medium, or with 10^{-5} *M* ABA. Some from each treatment were then processed for Gl II determination; others were cultured for a further 3 day period on fresh culture medium of the same composition as they had been on previously, or were transferred from basal medium to 10^{-5} *M* ABA medium or vice versa. On day 6 all embryos were processed for Gl II determination (Fig. 5). In the culture period day 3-6 embryos transferred from ABA to basal medium synthesized less Gl II than those that were continuously on ABA, and those transferred from basal to ABA medium synthesized more Gl II than those continuously on basal medium. These results indicate that ABA is continuously required to maintain the high rate of Gl II synthesis, and that embryos initially cultured in the absence of exogenously supplied ABA can subsequently respond to it by increasing their rate of Gl II synthesis.

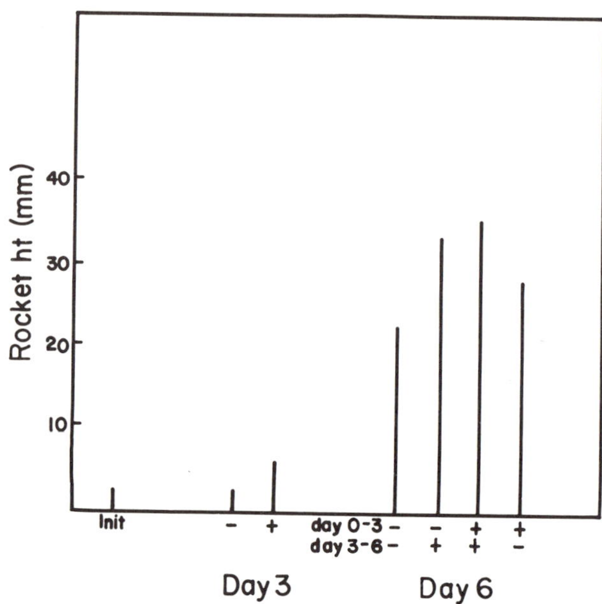

FIGURE 5. Glycoprotein II content of 9 day embryos removed from the seed (init) and those cultured for 3 days on basal medium (-) or on basal medium supplemented with 10^{-5} M ABA (+), and those cultured for 6 days on basal medium (--) or on medium supplemented with 10^{-5} M ABA (++), or transferred on day 3 from basal to 10^{-5} M ABA medium (-+), or transferred on day 3 from 10^{-5} M ABA to basal medium (+-).

This last observation provides a basis for interpreting the observation that the rate of protein accumulation in normal embryos increases rapidly after day 20, and that this is the time of rapid synthesis of Gl II in the embryo.

ABA STATUS OF THE DEVELOPING EMBRYO

The pattern of accumulation of ABA in embryos of *P. vulgaris* cv. Taylor's Horticultural has been investigated by Hsu (1979). He showed that prior to day 20 the ABA content of the embryo was low, and that there was a large increase in ABA content after day 20 (Fig. 6). ABA content was maximal on day 28 when the calculated concentration in the embryo was 1.2×10^{-5} *M*. Subsequently ABA declined and was undetectable in the mature seed.

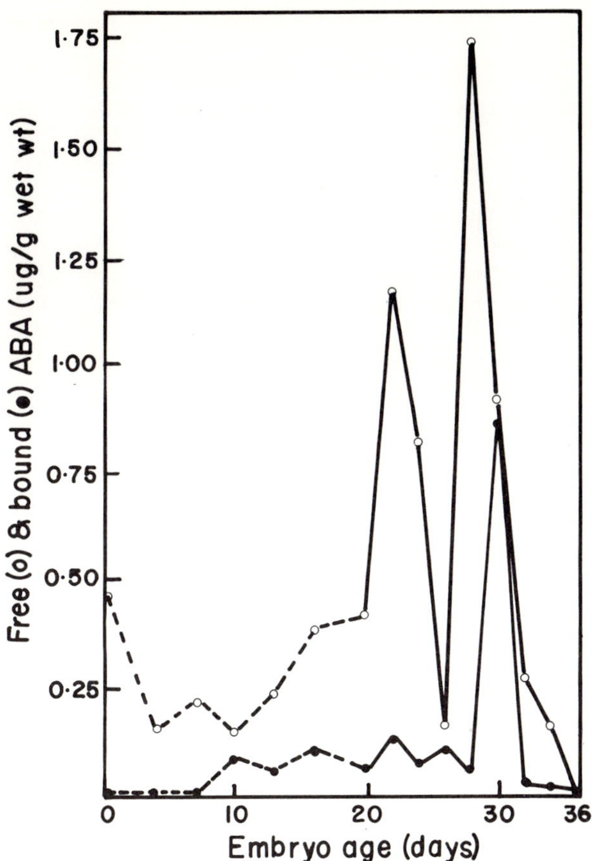

FIGURE 6. ABA content of developing embryos. Data points for day 0 and 4 are for whole fruit, those for day 7-16 are for whole seed (adapted from Hsu, (1979).

The correlation between the transition that occurs on day 20 from a condition of low ABA content and low rate of Gl II accumulation in the embryo to one of high ABA content and high rate of Gl II accumulation, and the experiments described here that show a direct stimulation of Gl II accumulation and synthesis by ABA strongly indicate that this hormone is implicated in regulation of Gl II content in the developing seed.

SUMMARY

Glycoprotein II, the major storage protein in embryos of *Phaseolus vulgaris*, is present in low amounts before day 20 in cv. Taylor's Horticultural, and accumulates rapidly thereafter. Day 20 also marks the transition between two different developmental phases of the embryo, characterized by differences in water content, protein, and abscisic acid. Glycoprotein II accumulation in 9 day embryos cultured aseptically was not stimulated by the hormones indole acetic acid, gibberellic acid, or kinetin, but was stimulated by concentrations of abscisic acid of 10^{-6} *M* or higher. Glycoprotein II synthesized by cultured embryos was stable until germination when it was metabolized. These results indicate that abscisic acid may be the seed hormone that regulates the synthesis of storage protein during development.

ACKNOWLEDGMENTS

We thank N. Gaedke and P. Naples for their expert technical assistance.

REFERENCES

Axelsen, N. H., Kroll, J., and Weeke, B. (1973). "Quantitative Immunoelectrophoresis." Univ. Oslo-Bergen-Tromso.

Bollini, R., and Chrispeels, M. J. (1978). Characterization and subcellular localization of vicilin and phytohemagglutinin, the two major reserve proteins of *Phaseolus vulgaris* L. *Planta 142,* 291-298.

Hartree, E. F. (1972). Determination of protein: A modification of the Lowry method that gives a linear photometric response. *Anal. Biochem. 48,* 422-427.

Hsu, F. C. (1979). Abscisic acid accumulation in developing seeds of *Phaseolus vulgaris*. *Plant Physiol. 63*,552-556.

Mante, S., and Boll, W. G. (1975). Comparison of growth and extracellular polysaccharide of cotyledon cell suspension cultures of bush bean (*Phaseolus vulgaris* cv. Contender) grown in coconut milk medium and synthetic medium. *Can. J. Bot. 53,* 1542-1548.

McLeester, R. C., Hall, T. C., Sun, S. M., and Bliss, F. A. (1973). Comparison of globulin proteins from *Phaseolus vulgaris* with those from *Vicia faba*. *Phytochem. 2,* 85-93.

Pusztai, A., and Watt, W. B. (1970). Glycoprotein II. The isolation and characterization of a major antigenic and non-haemagglutinating glycoprotein from *Phaseolus vulgaris*. *Biochim. Biophys. Acta 207,* 413-431.

Racusen, D., and Foote, M. (1971). The major glycoprotein in germinating bean seeds. *Can. J. Bot. 49,* 2107-2111.

Sun, S. M., Mutschler, M. A., Bliss, F. A., and Hall, T. C. (1978). Protein synthesis and accumulation in bean cotyledons during growth. *Plant Physiol. 61,* 918-923.

Walbot, V., Clutter, M., and Sussex, I. (1972). Reproductive development and embryogeny in *Phaseolus*. *Phytomorph. 22,* 59-68.

Seed Germination

PHYSIOLOGICAL AND OTHER ASPECTS OF SEED PRESERVATION

Louis N. Bass

USDA, SEA-AR
National Seed Storage Laboratory
Fort Collins, Colorado

Agriculture in the United States has, since the beginning, depended heavily upon plants introduced from other countries. Plant introduction was poorly organized and rarely funded until 1899 when an inventory of plant introductions was begun wherein each introduction received a number. Up to now, over 422,000 introductions have been numbered.

Unfortunately, initiation of the plant introduction inventory was not accompanied by an effective preservation system. New introductions were sent either to specialists in the Bureau of Plant Industry or to research workers in state experiment stations, none of whom had good seed storage facilities. Plant introductions that did not exhibit desirable attributes were either neglected until their viability was lost or discarded. Such losses led to repeated introductions of the same materials. It has become impossible to enter many fruitful geographic areas to reintroduce lost germplasm that might contribute greatly to our plant breeding programs.

THE NATIONAL SEED STORAGE LABORATORY

In 1944, the National Research Council recommended that the United States Department of Agriculture establish a facility for the preservation of valuable germplasm. After many years of planning and hard work by a special committee on the National Seed Storage Facility, Congress appropriated funds for construction and operation of the National Seed Storage Laboratory in 1956. The Laboratory, located on the campus of Colorado State University, Fort Collins, Colorado, opened in the fall of 1958.

ISBN 0-12-602050-7

It (Fig. 1) is a three-level building made of case reinforced concrete slabs. The cold storage rooms (Fig. 2) are well insulated with cork.

All refrigeration and air conditioning equipment is on the ground floor, as are the growth chamber room, the physiology research laboratory, and the control room. The administrative offices occupy the second level. The seed storage rooms and the germination laboratory (Fig. 3) are on the third level.

The ten storage rooms, seven of which are presently in use, are accessible from a common corridor, and have a capacity of about 180,000 pint cans. If necessary, the storage capacity can be greatly increased by changing from cans to flexible containers which can vary in size with sample volume.

In the storage rooms, the accessions are arranged in numbered steel trays, placed in numbered steel racks. Any risk of fire is practically eliminated, except for the cork insulation which is protected by a half inch of plaster.

The original plans called for all the storage chambers to be maintained at approximately 4°C and 35% relative humidity,

FIGURE 1. Plant physiologist Louis N. Bass (left), Director of the National Seed Storage Laboratory, and Botanist Dorris C. Clark in front of the USDA-SEA facility on the Colorado State University campus in Fort Collins. The Laboratory currently employs 16 scientists, technicians, and support personnel (0178X046-23).

FIGURE 2. The National Seed Storage Laboratory has ten storage rooms, of which four are already filled and three more are nearly filled. Here agronomist Phillip C. Stanwood selects a sample for a germination retest. As supervisor of the Laboratory's seed germination and storage programs, Dr. Stanwood is responsible for testing incoming seeds for germination percentage and then storing them according to individual temperature requirements. The storage rooms can be kept at temperatures that vary from just above freezing to as low as -12°C (1276X1561-24).

the humidity to be controlled by the reheat method. By the reheat method air is delivered from the cooling coil at a temperature 12 to 15° lower than the desired temperature, then reheated to the desired temperature. The 4°C temperature was selected as being suitable for most seeds for at least 10 years' storage. Three rooms, however, were modified and equipped so that a temperature of -10 to -12°C could be maintained. Other rooms are being modified to maintain cooler temperatures. For research purposes, three walk-in chambers were installed, each with three inner relative humidity chambers. These chambers have temperatures of 10, 21, and 32°C and relative humidities of 50, 70 and 90%, thus providing nine climatic analogs that

FIGURE 3. Germination tests indicate how well seeds retain their viability during long-term storage. Student assistant James Gill drops a random sampling of beans through a counting board onto a moist germination towel while technician Patricia Klein evaluates bean seedlings at the end of a germination test (0178X052-11).

range from the prevailing conditions in the northern states to those found in the deep south.

Only seeds are stored. They are accepted from all public agencies, seed companies, and individuals engaged in plant breeding or seed research. Obsolete varieties are accepted from anyone who has maintained them over the years. As much descriptive information as possible is obtained for each seed lot.

When received, each sample is given a serial number and a genus and species number. The seeds are then tested for viability. If the germination is satisfactory, the seeds are stored. For those kinds of seeds that can be dried to a low moisture content, a portion of each sample is dried to between 5 and 7% moisture content and stored at -10 to -12°C in hermetically sealed metal cans.

The germination laboratory has two large water-curtain room-type walk-in germinators, plus several small ones, that provide suitable temperature conditions for testing all kinds of seeds.

When the germination test shows a seed sample to be unsatisfactory for long-term storage, a new sample with higher germination is requested. The poor quality seed is stored temporarily until new seed is received.

When seeds are accepted for storage, the Laboratory has the responsibility for future maintenance, except when other arrangements are made at the time of acceptance.

Most seed lots are tested for germination every 5 years. When either germination or seed quantity drops to an unsafe level, the Laboratory arranges through contracts or some other method to produce a new generation with the same genetic composition as the original seed.

For most kinds of seed, between 10,000 and 20,000 seeds are requested. However, in the case of certain difficult-to-produce genetic materials, 500 seeds are acceptable, provided the donor assumes the responsibility for future increases.

All seeds accepted for storage in the National Seed Storage Laboratory become public property and are available for disbursement subject to certain restrictions. The Laboratory is not a seed distributing agency, but any bona fide research worker in the United States or its possessions can obtain nominal amounts of seed without charge, provided the requested germplasm is not available elsewhere.

All deposits and requests for seeds from foreign countries are handled through the Plant Introduction Officer, Germplasm Resources Laboratory, Science and Education Administration, USDA, Beltsville, Maryland. The National Seed Storage Laboratory does accept seeds from foreign countries and sends seeds to foreign scientists, but only with the approval of the Plant Introduction Officer.

The Laboratory also cooperates with the Food and Agricultural Organization of the United Nations (FAO) and the International Board of Plant Genetic Resources. All foreign proposals for storage are reviewed for approval by the SEA-AR Plant Germplasm Coordinating Committee. In making its decisions, the Committee is guided by recommendations from appropriate crop advisory committees. Acceptance for storage may require an exchange of letters between SEA-AR and the requesting agency or institution. Collections accepted for long-term storage (i.e., base collections) are accessioned and incorporated as an integral part of the Laboratory's inventory and hence the U. S. National Plant Germplasm System. Collections for temporary or emergency storage may be accepted, but under terms specified in an exchange of letters between SEA-AR and the requestor.

The Laboratory issues periodic inventories of the stocks held in permanent storage to inform research workers of materials available. Separate sections are provided for most crops. Inventories are available to scientists upon request. At present, more than 98,000 accessions are in storage.

The National Seed Storage Laboratory not only stores plant germplasm for posterity, but it also conducts research on ways to preserve seed viability.

GENETICS

The storage potential of seeds is influenced by inherent as well as external factors. For example, genetic differences between genera, species, and cultivars exist. Essentially, all seeds known to survive for 100 years or more belong to genera with hard, impermeable seed coats. Harrington (1972) listed *Albizia* as surviving for 147 years; *Cassia,* 158 years; *Goodia,* 105 years; and *Trifolium,* 100 years. All are Leguminosae, a plant family noted for species that produce hard seeds. However, barley, which does not have hard seeds, has been reported to have survived 123 years while sealed in a glass tube in a building stone in Nuremburg, Germany (Aufhammer and Simon, 1957). Seeds reported to have survived for over 500 years, such as *Canna* (Anonymous, 1968), *Lotus* (Ohga, 1923), and *Lupinus* (Porsild *et al.*, 1967) are hard-seeded. Not all Leguminosae species are long-lived; for example, the peanut is notoriously short-lived. Other short-lived kinds include onion, parsnip, and lettuce.

Shands *et al.* (1967) reported that 'Oderbrucker' outlived other barley cultivars when stored under the same conditions. Toole and Toole (1954) found that Black Valentine bean seeds lived longer in storage than did seeds of Brittle Wax. Significant cultivar differences in longevity of seeds of bean, pea, watermelon, cucumber, and sweet corn were reported by James *et al.* (1967).

PREHARVEST ENVIRONMENT

Austin (1972) reviewed the literature on the effect of preharvest environment on seed germination, but found no paper that related preharvest factors to seed storability. Mature seeds of normal size and appearance, free from mechanical injuries and microorganisms, that have not been subjected to temperature and moisture extremes during maturation, harvesting, and processing should store well. Conditions that affect any of these seed qualities can affect seed storability.

MacKay and Tonkin (1967) studied the effect of location of production on storability of seeds. They reported that Canadian-grown red clover seeds required 4 years to deteriorate to 80% germination, compared to 3 years for seeds grown in England and New Zealand. However, they did not establish that production conditions were the same; therefore, one cannot conclude that the location of production was entirely responsible for the differences in keeping quality.

Weather is probably the preharvest factor that has the greatest effect on seed viability. Dillman and Toole (1937) stored seeds of four flax cultivars grown under irrigation in California in 1929 and 1930. The 1930 seeds "showed marked weather injury." After 6 years of storage, seed of the four cultivars germinated 1, 4, 0, and 9%, whereas germination of the corresponding 1929 crop seeds was 94, 86, 87, and 94%. MacKay and Tonkin (1967) correlated the weather conditions during ripening and harvesting of barley, wheat, and oats grown in England with the number of years required for the seeds to deteriorate to either below 80% or below 50% germination. According to Riddell and Gries (1956) variations in growth of spring wheat from seeds of different ages was related to temperature during maturation rather than to age of seeds or storage conditions. Harrington and Thompson (1952) reported that location of production had a significant effect on the germination of lettuce seeds at temperatures between 24 and 30°C.

Moss *et al.* (1972) found that preharvest rains can cause wheat to germinate in the head, which, of course, reduces seed quality and storability.

Early sustained freezes can cause serious damage to corn of high moisture content. The degree of damage is influenced by the temperature reached, duration of low temperature, and moisture content of the seeds.

SEED STRUCTURES

Seed structures such as the presence or absence of glumes (lemma and palea) in grasses markedly influence seed lifespan. Haferkamp *et al.* (1953) found that aged seeds of barley and Red Winter Speltz wheat with "hulls" retained viability better than seeds that had been threshed and stored. Hulls and chaff had an inhibitory effect on the growth of molds, suggesting that suppression of mold growth increased the lifespan of cereal seeds. Lakon (1954) showed that oat and timothy seeds had a longer lifespan when stored with the glumes intact than when stored hulled.

Esbo (1954) reported that the viability of hulled timothy seeds declined 16% during the first year, but seeds with the hulls intact showed no significant loss in viability until the third year. Field emergence of unhulled seeds was 6-14% higher than was emergence of hulled seeds. Stevens (1935) stored unhulled and hulled timothy seeds under the same conditions for 11 years. The germination of unhulled seeds declined from 98 to 52%, and germination of hulled seeds declined from 97 to 16%.

Roberts (1972) discussed the protection from mechanical damage during harvesting, handling, and processing provided by seed shape. He also observed that small seeds usually escape injury, whereas large seeds frequently suffer extensive damage. Size, arrangement of essential seed structures, and seed composition were contributing factors. Bean and lima bean are highly susceptible to damage, whereas corn is moderately susceptible. According to Roberts, a spherical shape offered more protection than a flat or irregular shape. He observed that the embryonic root tips of onion extend beyond the body of the seed, a condition conducive to mechanical injury.

SEED COMPOSITION

Available literature does not relate seed composition or biochemistry to lifespan. Some relationship may exist, but more definite information is needed before this subject can be reasonably discussed. Chemical, physical, and nutritional changes during storage are discussed by Zeleny (1954) and biochemical indices of deteriorating seeds are discussed by Abdul-Baki and Anderson (1972).

HARD SEEDS

Much has been written about hard seeds (seeds which are impermeable to water within the usual laboratory germination test period), but little information is directly related to seed storage. Although hard seeds are usually associated with species of the legume family, they are frequently found in okra and hollyhock and infrequently in cotton in the mallow family (Malvaceae). They are also found in catnip in the mint family (Labiateae); cranesbill in the geranium family (Geraniaceae); canna in the Cannaceae; *Ipomoeae, Convulvulus,* and *Cuscuta* in the morning-glory family (Convulvulaceae); and Indian lotus in the Nymphaeaceae. It is possible that impermeable seeds are also produced by a number of species in other families, especially native wild plants. Under certain environmental conditions, the hard seed content of some seed lots increases during storage. Generally, a warm, dry atmosphere induces hard seed formation and a cool, moist atmosphere favors a low hard seed population. In many species, hard seeds have a longer lifespan than permeable seeds. This is an advantage to the survival of the species, but it causes problems in cultural practices. Storage of seeds at intermediate or high relative humidities to minimize the percentage of hard seeds in a lot contributes to more rapid deterioration of germination capacity. According to the Association of Official Seed Analysts (AOSA) Rules for Testing Seeds (Anonymous, 1970), hard seeds may be encountered in some seed lots of 34 genera of cultivated plants.

SEED MATURITY

Maturity of seeds at harvest also affects seed longevity. Scientists have regarded seed maturity as being that stage of maturation at which maximum dry weight has been attained (Harrington, 1972; Roberts, 1972). Because many crop species flower and mature seeds over a period of several days or even weeks, it is important to know at what stage of maturity the seeds should be harvested. McAlister (1943) harvested seeds of three species of *Agropyron,* three species of *Bromus, Elymus glaucus,* and *Stipa viridula* in the premilk, milk, dough, and mature stages and stored them at temperatures ranging between 15 and 23°C and moisture contents of 7-9%. After storage of 4-5 months, viability of the seeds harvested in the premilk and milk stages was inferior to the viability of seeds harvested in the dough and mature stages. Mature Kentucky bluegrass seeds remained viable longer than immature seeds when both were stored under the same conditions (Bass, 1965). Mature seeds that germinated 93% at harvest germinated 53% after 93 months at 2°C and 70% relative humidity, but immature seeds that germinated 88% when harvested germinated only 15% after storage. Mature seeds stored at 32°C and 15% relative humidity germinated 81%, and immature seeds germinated 59%.

Rate of drying had little effect upon the response of mature and immature Kentucky bluegrass seeds to storage conditions (Bass, 1965). Relative humidity had a greater effect than temperature did on the longevity of these seeds. Seeds held at 32°C/15% RH had fair to good germination after 93 months of storage, whereas similar seeds at 2°C/70% RH had poor germination (Bass, 1965). Similar results have been reported by numerous other workers: Esbo (1959) for *Phleum pratense*, Hermann and Hermann (1939) for *Agropyron cristatum*, Griffith and Harrison (1954) for *Phalaris arundinacea*, Jensen and Jorgensen (1969) for *Festuca pratensis*, and others.

Eguchi and Yamada (1958) harvested seeds of cabbage, carrot, Chinese cabbage, cucumber, edible burdock, eggplant, Japanese radish, pumpkin, tomato, watermelon, and Welsh onion at several stages of maturity and stored them for 2 years. Eight kinds showed marked losses in viability of immature seeds compared to mature seeds. However, tomato, pumpkin, and Chinese cabbage showed little or no difference in longevity with different maturities.

Seed size or weight had no effect on lifespan of seeds. However, numerous studies have shown the superiority of heavy, mature seeds over light, immature seeds in germination, vigor, and yield. Relatively few exceptions have been noted. Good reviews are provided by Black (1959) and Austin (1972). Austin discussed seed maturity and seed size jointly because seed size is determined or mediated to some extent by maturity. His discussion supports the view that the same or similar problems beset immature and small seeds in storage.

SEED DORMANCY

Dormancy of freshly harvested seeds may be found in practically all groups of plants. In most crop species, dormancy is dissipated within a few to several months if seeds are stored at ambient temperatures and relative humidities or under controlled atmospheres, provided the temperature is held above freezing. The best known method of maintaining dormancy in seeds is storage at subfreezing temperatures. Owen (1956) and Koller (1972) have reviewed pertinent literature on the existence of seed dormancy and its dissipation with time. Roberts

(1972) reviewed pertinent literature and concluded that available evidence is not sufficient to establish even a casual relationship between dormancy and lifespan.

SEED MOISTURE CONTENT

Seed moisture content and storage temperature are very important factors in seed longevity, with seed moisture content usually regarded as most influential. Barton (1961) regarded seed moisture content and temperature as vitally important factors in seed deterioration. Because these two factors are so important and interrelated, it is difficult to discuss them separately. It is a well-established fact that within certain limits, seed deterioration increases as seed moisture content is increased. The literature contains much substantiating evidence gained from research conducted under various circumstances and conditions.

Bass (1953) found that the loss of viability of freshly harvested Kentucky bluegrass seeds was correlated with seed moisture content and length of time at a given temperature. Seeds with 54% moisture lost 20% germination during 45 hr at 30°C, but seeds with 44% moisture showed no loss of viability during 36 hr at 45°C. Seeds with 22 and 11% moisture content showed essentially no loss in viability during 45 hr at 50°C. McNeal and York (1964) concluded that sorghum for seed should be harvested at 20% seed moisture or less and dried promptly to about 11% moisture content. The drying temperature should not exceed 43.3°C for very moist seeds and 45.4°C for low moisture seeds.

Many kinds of seeds can be dried to 6% moisture content without damage, but some kinds are injured by drying to lower moisture levels. However, seeds of numerous crop species can be dried to 2 to 3% moisture content, provided other factors associated with drying do not cause injury. Drying below 3 to 4% moisture content is not recommended for seeds in commerce because too rapid rehydration may cause problems and very dry seeds are easily damaged on impact. There is extensive literature on the influence of temperature and seed moisture content on the storage life of seeds of many crop species.

WATER ABSORPTION AND RETENTION

Because seed deterioration is influenced by moisture content, it is important to know what factors affect water absorption and retention. The thickness, structure, and chemical composition of the seedcoat influence the rate of water absorption by seeds; in the case of hard seeds, the seedcoat restricts or limits total water uptake. The chemical composition of the seed influences the amount of water absorbed and held. Of the various seed constituents proteins are the most hygroscopic, carbohydrates are somewhat less so; and the lipids are hydrophobic. Thus, seeds which contain relatively high percentages of carbohydrates and/or proteins have moisture contents of about 13 to 15% at 25°C and 75% relative humidity, whereas seeds that are rich in oil have moisture contents of about 9 to 11% at the same temperature and relative humidity.

The most common method of determining seed moisture content is by heating the seeds in a forced air oven at a given temperature for a specified time or until constant weight is obtained. The loss of water represents the moisture content of the seeds. The percentage of moisture may be calculated by dividing the weight loss by either the wet weight or dry weight. Moisture percentages based on sample dry weight are frequently used in research, and percentages based on sample wet weight are usually used for commercial purposes. Seedsmen and seed testing laboratories (Association of Official Seed Analysts, Society for Commercial Seed Technologists, and the International Seed Testing Association) use the wet weight method of determining seed moisture content. Seed moisture percentages calculated by the dry weight method are usually higher than those calculated by the wet weight method.

Relative humidity expresses the amount of moisture actually in the air as a percentage of the amount of moisture the air is capable of holding at a specified temperature. The importance of seed moisture content in seed deterioration cannot be overemphasized. Under all storage conditions seed moisture content comes to equilibrium with the moisture in the surrounding air. The time required for seeds to reach equilibrium is regulated by the time required for moisture to penetrate the seed coat and the time required for moisture transfer within the seed.

ADVERSE EFFECTS OF LOW MOISTURE

Although it is very important to reduce seed moisture content to a safe level for storage, one must also be aware of the possible adverse effects of low moisture content, as very dry seeds are susceptible to mechanical breakage and related injuries. Seed structure and seed resistance to removal from the pod or from the mother plant, as in the grasses, influence the amount of damage. Beattie and Boswell (1939) and Moore (1972) have shown that damaged seeds do not store as well as intact seeds and that fungi enter the seeds through cracks in the seed coat. In natural deterioration in spinach seeds the percentage of abnormal seedlings increases with time; the percentage of normal seedlings decreases. Thus seedling abnormalities result not only from mechanical damage but also from natural aging. Threshing or combining produces breaks, cracks, bruises, and abrasions in seeds, which in turn produces seedlings of questionable value. There is good evidence that damaged seeds of small-seeded legumes do not survive as long as nondamaged seeds. Battle (1948) found that alfalfa seeds scarified with sandpaper were dead after 14 years' storage, whereas unscarified seeds germinated 23%. Similar results were reported by Graber (1922), Stevens (1935), and Brett (1952). Oathout (1928) and Mamicpic and Caldwell (1963) showed that damaged soybean seeds lose viability more quickly than nondamaged seeds, and Blackstone *et al.* (1954) reported similar results for peanuts. Studies on seed injury resulting from mechanical harvesting, threshing, combining, and handling have seldom been concerned with storage. However, vigor at the time of storage influences seed storage life.

TEMPERATURE

Temperature is the most important factor in seed storage after moisture content. Within limits, the storage life of seeds decreases as temperature increases. Toole and Toole (1946) stored seeds of Mammoth Yellow and Otootan soybeans at five temperatures and three moisture contents for a period of 10 years. There was no significant loss in germination of 8 to 9% moisture seed stored at 20, 10, 2, and -10°C. In field plantings, no apparent difference in vigor of growth was observed between 10-year-old seeds that had shown no loss in germination and 1-year-old seeds of both cultivars.

The superiority of subfreezing temperature to higher temperatures for seed storage has been well established. However, Weibull (1952, 1955) found that seeds of a few species did not benefit by storage at a subfreezing temperature. The temperature of -20°C was unfavorable for storage of parsley, snapdragon, fern asparagus, hybrid petunia, and pansy seeds.

The effects of ultracold temperatures have also been studied. As early as 1895, de Candolle and Pictet reported that they had frozen seeds for 2 hr at -180°C and -100°C for 4 days without adversely affecting germination. Brown and Escombe (1897-1898) exposed seeds of 12 species (moisture content 10-12%) to -183°C for 110 hr without damage. In 1909, White reported that exposure to -200°C for 1.5 days caused severe injury to apple, hemp, lobelia, parsnip, and parsley seeds; slight injury to turnip, bean, mustard, pea, radish, and sunflower seeds; and no injury to cress and castorbean seeds. Barley, corn, oat, rye, and wheat seeds showed no decrease in germination after 2 to 3 days at -200°C. Becquerel (1953) dried clover, alfalfa, petunia, and tobacco seeds, then held them for 2 hr at -272.9°C without loss of germination. In 1934 Lipman and Lewis stored seeds of sugarcane, spinach, cucumber, sugarbeet, buckwheat, barley, purple vetch, oat, onion, mustard, and sweet clover at -196°C for 30 days, and seeds of pea, corn, squash, alfalfa, and sunflower for 60 days without obtaining a decline in either germination or vigor. In 1936, Lipman stored vetch, wheat, barley, tobacco, flax, buckwheat, milo, spinach, and sweet clover seeds at -272°C for several hours without loss of germination. Seeds of rice, winter wheat, soybean, alfalfa, and ryegrass with seed moisture contents ranging from 5 to 26% (dry weight basis) were exposed to liquid nitrogen (LN_2), then rewarmed at different rates (Sakai and Noshiro, 1970). Seeds below 8% moisture were not damaged by LN_2 temperature, and rewarming rate was not critical.

Because LN_2 may be practical for germplasm preservation, long-term studies on its use have begun at the National Seed Storage Laboratory. To date, seeds of more than 100 species have been stored in LN_2 (-196°C) for 7 to 180 days. For most kinds of seeds, germination percentage was not decreased, even after 180 days' storage. Studies on some species will last for at least 50 years if the seeds remain viable that long. Should germination continue to remain unchanged for several years, the interval between tests will be increased and the termination date for the study will be changed accordingly. It is realized that LN_2 storage will probably never be used by the seed trade, except possibly for certain high-priced small-seeded types. However, LN_2 storage may have some advantages for germplasm preservation if only small samples are stored. If, as is anticipated, LN_2 storage does extend storage life, the savings in both time and money spent on seed increases can be substantial.

Research on the effects of high temperatures on seeds has shown that seed vitality and vigor are usually reduced as temperature is increased, as time at temperatures is increased, and as seed moisture content is increased. At a given high temperature, damage is diminished as moisture content is decreased. Consideration of temperature, length of temperature exposure, and moisture content of the seeds is essential when drying seeds for storage.

Most research on temperature effects has shown that as temperature is increased and the relative humidity held constant, seed moisture content is decreased. Toole *et al.* (1948) stored seeds of 15 vegetable species at three relative humidities and three temperatures. For seeds of all species, moisture content was lower at 22°C and 27°C than at 11°C. However, without exception, moisture contents at 27°C were higher than those at 22°C. Although seed moisture content increases with an increase in relative humidity, changes in temperature have little effect on seed moisture content.

Temperature also influences the rate of water absorption. Dillman (1930) reported that the rate of moisture absorption by dry seeds of wheat, corn, and flax was twice as rapid at 30°C as at 10°C but was the same at 40°C as at 30°C. The rate of moisture movement is determined by the difference between the vapor pressure in the seeds and that of the surrounding atmosphere.

Moisture uptake or loss is rapid at first, but slows as equilibrium with the surrounding atmosphere is approached. At high relative humidities, changes in sample weight may be influenced by changes in water content of the seeds, by growth of microorganisms, and by respiration.

Because movement of moisture into or out of a bulk of seeds is a slow process, large bulks of seeds can be stored safely. Moisture is distributed within a seed much more rapidly than through a mass of seeds.

Over the years numerous studies have been conducted on the effects of controlled storage conditions. In some studies, only temperature was controlled; in others either the gas surrounding the seeds or various combinations of temperature, relative humidity, and surrounding gas were controlled.

CONTROLLED ATMOSPHERE STORAGE

Controlled atmosphere storage studies have utilized a wide variety of seeds, although the more extensive studies have been limited to relatively few kinds. It is not possible to review all the studies that have been made on controlled atmosphere storage; therefore this discussion is limited principally to studies that have been or are being conducted in the National Seed Storage Laboratory.

It is well known that suitable storage conditions must be used if seeds are to be held in good viable condition for several years. It is frequently said that seeds must be held at a low temperature and a low relative humidity in order to keep them viable for a period of years. However, it is not known how low the temperature and/or relative humidity must be in order to assure good viability for a specific length of time.

According to Harrington's rule of thumb, (*a*) the sum of the percentage of relative humidity plus the temperature in degrees F should not exceed 100 for storage up to 5 years; and (*b*) the length of time seeds may be stored without a significant decline in viability doubles for each 1% drop in seed moisture and for each 5°C drop in temperature. With these two rules of thumb in mind, let us take a look at some storage data.

Hand-shelled Spanette peanut seeds were stored 5 years at temperature °C/RH combinations of 10/50, 10/70, 21/50, 21/70, and 32/50. The 10/50 combination, according to Harrington's rules, totaled 100. The peanut seeds germinated 98% when first stored and 98% after 5 years at 10/50. The other temperature/RH combinations exceed 100; however, they provided some interesting information. Theoretically, 10/70 and 21/50 as equivalents should have given the same results, but they did not. Seeds that germinated 98% when stored germinated 91% after 5 years at 10/70; and similar seeds stored at 21/50 germinated only 85% after 5 years. A look at the results for seeds at 21/70 and 32/50, which have the same total of temperature and relative humidity, shows that germination declined more rapidly at 21/70 than at 32/50. In 2 years, the germination of Spanette peanut seeds dropped from 98% to 79% at 21/70 but only to 86% at 32/50. Equilibrium moisture content was 7.8% at 21/70 and 5.2% at 32/50, which probably accounted for the difference in rate of viability loss (Bass, 1968).

All or nearly all seeds of winged *Dimorphotheca sinuata* died within 3 months when stored at 32°C/90% RH and 21/90; within 6 months at 32/70; within 12 months at 10/90; and within 24 months at 21/70, 32/50, and 21/50. Seeds stored at 10/70 retained one-half of their initial viability after 42 months, whereas seeds at 10/50 showed no loss in viability after 134 months (Bass *et al.*,1967 and unpublished data).

Seeds of bean, pea, watermelon, cucumber, tomato, and sweet corn cultivars from 3 crop years were stored in all combinations of 10, 21, and 32°C, and 50, 70, and 90% RH for up to 5 years (James *et al.*, 1967). Significant effects of year of production on viability maintenance were evident in seeds stored under adverse conditions, 32/90 and 32/70. Only minor differences were observed for seeds stored at 10/70 and 10/50.

Seeds of most cultivars used in this study showed no significant loss of viability during 9 and 10 years of storage at 10/50. However, a few cultivars of most kinds of seeds did show a significant viability decline. Such losses point up the need for extreme care in making broad, general statements about seed storage requirements.

Comparison of three reciprocal pairs of temperature/relative humidity combinations, 32/70 versus 21/90; 32/50 versus 10/90, and 21/50 versus 10/70, showed that all seeds except sweet corn retained their viability better at 32/70 than at 21/90. The 32/50 combination provided better storage conditions for watermelon and cucumber seeds, but the 10/90 combination provided better conditions for bean and sweet corn seeds. The difference between the 10 and 21°C combinations were significant for seeds of pea and watermelon (James *et al.*, 1967).

Relative humidity had a marked effect upon longevity of green and bleached lima bean seeds stored at 21°C (Bass *et al.*,

1970). At the end of 36 months, no significant loss of viability had occurred in either green or bleached seeds at 50% RH. At 70% RH the bleached seeds had lost all viability after 18 months, but the green seeds showed no significant loss of viability. At 90% RH the bleached seeds were dead after 3 months, and the green seeds had lost about one-half of their initial viability.

Reed canarygrass seeds sometimes lose up to one-half their viability between receipt in the seed company warehouse and final processing. Such losses can be largely overcome by controlling the temperature and/or relative humidity of the storage area. In a study on seeds harvested from the same fields in three consecutive years, the seeds from no one crop year consistently, under all storage conditions, retained their germination better than did seeds from the two other crop years. Seeds stored at 32, 21, and 10°C under various RH conditions did not attain the same moisture contents; neither did seeds stored at 90, 70, and 50% RH at various temperatures.

The rapidity of loss of viability of reed canarygrass seeds increased as the sum of the RH and temperature (degrees F) increased above 100. The rate of viability loss was most rapid when the temperature was higher than the relative humidity. Combinations of temperature (degrees F) and relative humidity that total more than 100 are not satisfactory for long-term storage of reed canarygrass seeds in open or porous containers. However, combinations totaling 120 may be safe for 1 to 3 years, providing temperature contributes less than one-half the total (Bass, 1967).

An abnormality of lettuce seeds called physiological necrosis, red cotyledons, or spotted cotyledons has been recognized for many years, but its cause is unknown. This abnormality is seldom present in freshly harvested seeds; consequently, it is assumed to be caused by aging. However, portions of the same seed lot stored under different conditions do not develop necrotic seeds either simultaneously or in equal numbers. For example, seeds of Imperial 456 and Imperial 44 lettuce developed 95% necrotic seeds in 5 years at 10°C/50% RH. At 5°C/40% RH, 99% of the seeds were necrotic after 8.5 years, but at -12°C/70% RH seeds of both cultivars were free of necrosis after 8.5 years (Bass, 1970).

PROTECTIVE PACKAGING

It is possible to control the atmosphere surrounding stored seeds by special packaging. Hermetically sealed metal and glass containers are impervious to moisture vapor and gases. Containers made of flexible packaging materials resist transmission of moisture vapor and gases only to the extent that special barrier properties are built into them.

In impervious packages the relative humidity of the atmosphere is determined primarily by seed moisture content, and any change is limited to the small effect of temperature. The relative humidity in packages made of materials with limited permeability is determined by seed moisture content, temperature, and relative humidity of the storage area, permeability of the packaging material, and size of the package.

Seeds in packages that are not completely impervious to moisture vapor may gain or lose moisture with time. The direction, rate, and amount of change are controlled by the temperature and relative humidity of the storage area, moisture vapor transmission rate of the packaging material, equilibrium moisture content of the seeds for the surrounding temperature and relative humidity, and the ratio of surface area of the seeds to the surface area of the package. Because small packages contain fewer seeds per unit area of package surface than do large packages, seeds in small packages gain or lose moisture faster than seeds in large packages of the same material held under the same temperature and relative humidity. Thus, it is essential to use a good moisture-barrier material for small packages.

Many kinds of seeds do not require special moisture protection during the first winter after production if they are held in the area where they were produced or under similar climatic conditions. However, carry-over seeds often require drying and packaging in moisture-barrier containers (Bass *et al.*, 1961).

Before the wide variety of plastic materials and laminates available today, seeds were sealed in tin cans or glass jars and vials. However, some flexible materials, when properly sealed, provide essentially the same moisture protection as a sealed metal can.

Hemp seeds, which contained 9.5% moisture or less when sealed in metal cans, retained essentially their full viability for up to 15 years at 10, 0, and -10°C, except that the 9.5% moisture seeds held at 10°C lost viability slowly during the first 6.5 years of storage and very rapidly thereafter. Kenaf seeds responded similarly (Clark *et al.*, 1963).

Reed canarygrass seeds containing approximately 4, 7, and 10% moisture when sealed in metal cans showed no loss of viability during 12 years at temperatures of 10°C and lower, except that the 10% moisture seeds at 4°C declined about 10%. Seeds with 4% moisture content showed no loss of germination at 10°C but lost 29% at 21°C (germination, 57%) and almost all viability at 32°C. The seeds with 7% moisture content germinated only 35% after 6 years at 21°C and only 9% at 32°C. The seeds with 10% moisture at 32°C germinated only 4% after 1 year, whereas seeds at 21°C germinated 41% after 4 years and 5% after 6 years (Bass, unpublished data).

Studies to determine the value of thermoplastic materials as moisture-barrier containers for seeds have shown that thin gauges of plastic films, polyethylene, polyester, and cellophane are little better than paper as moisture-barrier materials. Some moisture protection is provided by asphalt laminated multiwall, polyethylene laminated multiwall, and thicker gauges (5 to 10 mil) of polyethylene bags. Good moisture protection is afforded only by laminated materials which include an aluminum layer at least .35 mil thick. Such laminates when properly sealed provide moisture protection essentially as good as that provided by a hermetically sealed metal can. For example, safflower seeds that contained 4.2% moisture when packaged in containers made of 1 mil acetate/1 mil foil/heat seal coating still contained only 4.2% moisture after 11 years of storage at 10°C/70% RH; 4.6% moisture after 11 years at 10/90; and 4.3% moisture at 21/90. Seeds at 32/70 contained 4.4% moisture after 9 years but had lost all viability after 4.5 years. Seeds at 21/90, 10/90, and 10/70 still germinated above 60% after 11 years. Similar seeds sealed in bags made of 25# Kraft/7#PE/.35 mil foil/15#PE contained 4.8% moisture after 11 years at 10/70 and 5.3% moisture at 10/90. The seeds stored at 10/70 germinated 61%, and those stored at 10/90 germinated 72%. Initial germination was 96%. Seeds stored in this material at 32/90 had a moisture content of 4.1% after 5 years but germinated only 17%. Moisture content after 7 years was 4.8% and only three very weak seedlings were obtained.

Safflower seeds stored at 4.2% moisture in 2 mil metalized polyester packages at 10/70 and 10/90 contained 5.0 and 5.1% moisture and germinated 89 and 95% respectively, after 11 years of storage. Similar seeds held at 32/70 contained 4.8% moisture after 8 years but lost all viability after storage for 6.5 years (Bass and Clark, 1974).

MOISTUREPROOF STORAGE

To further illustrate the relationship between storage temperature, seed moisture content, and seed longevity, let us consider briefly a study in progress in the National Seed Storage Laboratory. Sorghum, crimson clover, safflower, sesame, and lettuce seeds with 4, 7, and 10% moisture were sealed in metal cans and stored at -12, -1, 10, 21, and 32°C.

Sorghum seeds of all moisture levels retained essentially their initial germination after 8 years at -12, -1, and 10°C. Seeds with 4 and 7% moisture showed some decline at 21°C, and at 32°C they lost about one-third of their initial germination.

Germination of seeds with 10% moisture dropped from 91 to 43% in 5 years and to 0 in 8 years.

Crimson clover seeds with 4 and 7% moisture retained their germination well for 8 years at all temperatures. Seeds with 7% moisture at 21°C showed the greatest amount of viability loss. Seeds with 10% moisture showed a marked decline in germination between the 5 year and 8 year tests for seeds at 10°C; they showed a gradual decline to 5% germination at 21°C. Germination of seeds at 32°C dropped to 41% in one year, and by the end of the second year only a few (2 to 6%) hard seeds survived.

Safflower seeds sealed at 4% moisture germinated about 90% after 8 years of storage at all temperatures included in the study. Seeds with 9% moisture retained good germination at 10, -1, and -12°C but dropped to 3% germination in 3 years at 21°C and to 0 in 1 year at 32°C. Seeds with 10% moisture germinated about 80% after 8 years at -12°C, 70% at -1°C and dropped to 17% at 10°C after 2 years. At 21 and 32°C, seeds with 10% moisture content died in less than 1 year.

Sealed sesame seeds with 4% moisture retained essentially full viability for 8 years at all five temperatures. Seeds with 7% moisture content stored well at -12 and -1°C, but at 10°C germination declined to zero in 8 years. Seeds at 32°C died in less than 1 year, and those at 21°C died in less than 2 years. Seeds at 10% moisture did not store well at any temperature.

Lettuce seeds with 4% moisture germinated 90% or higher after 8 years of storage at all temperatures except 32°C. Seeds held at 32°C produced 36% normal seedlings and 44% abnormal seedlings. Seeds with both 7 and 10% moisture retained essentially full initial viability at -12°C and -1°C. Germination of seeds with 7% moisture at 10°C declined from 94% after 5 years to 1% after 8 years of storage. At 10°C the 10% moisture seeds germinated 80% after 3 years and only 4% after 5 years of storage. At 21°C seeds with 7% moisture germinated 92% after 3 years and 4% after 4 years, whereas the seeds with 10% moisture germinated only 11% after 1 year of storage. At 32°C all seeds with 7 and 10% moisture died in less than 1 year.

This study also introduced various atmospheres into the cans before sealing. Sorghum seeds with 4% moisture sealed in air and nitrogen germinated significantly higher than those sealed in air, but not higher than those sealed in other atmospheres. Seed with 10% moisture content were nearly all dead; however, a few seedlings were obtained from the seeds stored in a partial vacuum and argon, the atmospheres that gave the best results with seeds with 4% moisture.

Crimson clover seeds with 4 and 7% moisture were not significantly different. For safflower, sesame, and lettuce stored at 32°C, only seeds with 4% moisture survived for 8 years. Differences among atmospheres were not significant for any of the three kinds of seeds, except that the lettuce seeds sealed in air germinated significantly lower than did the seeds in all the other atmospheres. The very low germination of lettuce seeds in air may have resulted from a poor seal. Because of the limited supply of samples, a retest from another can was not feasible. The next test may show whether a poor seal was involved.

Results of this study suggest that for short-term storage there is no advantage in using either a partial vacuum or a gas other than air in sealed containers. In fact, seed moisture content has more effect upon longevity of seeds in sealed containers than does the composition of the surrounding atmosphere. However, vacuum or gas sealing of some kinds of seeds may have a long range advantage at certain moisture contents held at various temperatures indefinitely (Bass *et al.*, 1962, 1963a and b, Bass 1973; Justice and Bass, 1978, and unpublished data).

SUMMARY

Various factors, such as maturity at harvest, mechanical damage, and preharvest weather conditions affect the longevity of seed lots regardless of the storage conditions. Cultivar differences in longevity under specific storage conditions do exist. It cannot be stated categorically that seeds of all cultivars of a given kind (such as corn or beans) retain their viability longest under a particular set of storage conditions.

The lifespan of many kinds of seeds can be extended either by controlling the relative humidity and temperature of their storage area or by predrying and packaging in moistureproof containers. Whether in porous or moisture-barrier containers, storage at a subfreezing temperature extends the longevity of many kinds of seeds.

For short and intermediate storage periods, sealing in an atmosphere other than air has no advantage. However, there may be a long-term advantage for some kinds of seeds under certain storage conditions. For sealed storage it is essential that seed moisture content be at a level safe for the highest temperature to which the seeds might be subjected.

With all the knowledge available on seed preservation, the precise conditions required for maximum storage life of specific kinds of seeds are not known.

REFERENCES

Abdul-Baki, A. A., and Anderson, J. D. (1972). Physiological and biochemical deterioration of seeds. *In* "Seed Biology." Vol 2, (T. T. Kozlowski, ed.), pp. 283-315. New York.

Anonymous (1968). 550-year old seed sprouts. *Sci. News 94,* 367.

Anonymous (1970). "Association of official seed analysts rules for testing seeds." *Proc. Assoc. Off. Seed Anal. 60.*

Aufhammer, G., and Simon, U. (1957). Die samen landwirtschaftlicher kulturpflanzen im grundstein des chamaligen nürnberger stadttheaters und ihre keimfahigkeit. *Z. Acker-u Pflbau. 103,* 454-472.

Austin, R. B. (1972). Effects of environment before harvesting on viability. *In* "Viability of Seeds" (E. H. Roberts, ed.), pp. 114-149. London.

Barton, L. V. (1961). "Seed Preservation and Longevity." London and New York.

Bass, L. N. (1953). Relationships of temperature, time and moisture content to the viability of seeds of Kentucky bluegrass. *Iowa Acad. Sci. 60,* 86-88.

Bass, L. N. (1965). Effect of maturity, drying rate and storage conditions on longevity of Kentucky bluegrass seed. *Proc. Assoc. Off. Seed Anal. 55,* 43-46.

Bass, L. N. (1967). Response of reed canarygrass (*Phalaris arundinacea* L.) seeds to storage conditions. *Proc. Assoc. Off. Seed Anal. 57,* 124-219.

Bass, L. N. (1968). Effects of temperature, relative humidity and protective packaging on longevity of peanut seeds. *Proc. Assoc. Off. Seed Anal. 58,* 58-62.

Bass, L. N. (1970). Prevention of physiological necrosis (red cotyledons) in lettuce seeds (*Lactuca sativa* L.). *J. Amer. Soc. Hort Sci. 95,* 550-553.

Bass, L. N. (1973). Controlled atmosphere and seed storage. *Seed Sci. and Technol. 1,* 463-492.

Bass, L. N., and Clark, D. C. (1974). Effects of storage conditions, packaging materials, and seed moisture content on longevity of safflower seeds. *Proc. Assoc. Off. Seed Anal. 64,* 120-128.

Bass, L. N., Ching, T. M., and Winter, F. L. (1961). Packages that protect seeds. *U. S. Dept. Agr. Ybk. 1961,* 330-338.

Bass, L. N., Clark, D. C., and James, E. (1962). Vacuum and inert-gas storage of lettuce seed. *Proc. Assoc. Off. Seed Anal. 52,* 116, 122.

Bass, L. N., Clark, D. C., and James, E. (1963a). Vacuum and inert-gas storage of safflower and sesame seeds. *Crop Sci. 3,* 237-240.

Bass, L. N., Clark, D. C., and James, E. (1963b). Vacuum and inert-gas storage of crimson clover and sorghum seeds. *Crop Sci. 3,* 425-428.

Bass, L. N., James, E., and Clark, D. C. (1970). Storage response of green and bleached lima beans (*Phaseolus lunatus* L.). *Hortscience 5,* 170-171.

Bass, L. N., Toy, S. J., Sayers, L., and Clark, D. C. (1967). Storage of *Dimorphotheca sinuata* D. C. and *Osteospermum ecklonis* Norl. Seed. *Proc. Assoc. Off. Seed Anal. 57,* 67-70.

Battle, W. R. (1948). Effect of scarification on longevity of alfalfa seed. *Amer. Soc. Agron. J. 40,* 758-759.

Beattie, J. H., and Boswell, V. R. (1939). Longevity of onion seed in relation to storage conditions. *U.S. Dept Agr. Cir. 512.*

Becquerel, P. (1953). La suspension de la vie aux confins du zèro absoluet ses consèquences. "Extrait des Actes due Congres de Luxembourg," 72nd Session Assoc. Franc. Avanc. Sci., pp. 487-491.

Black, J. N. (1959). Seed size in herbage legumes. *Herb. Abs. 29,* 235-241.

Blackstone, J. H., Ward, H. S., Jr., Butt, J. L., Reed, I. F., and McCreery, W. F. (1954). Factors affecting germination of runner peanuts. *Ala. Agr. Expt. Sta. Bul. 289.*

Brett, C. C. (1952). Factors affecting the viability of grass and legume seed in storage and during shipment. *Int. Grassland Conf. 6,* 878-884.

Brown, H. T., and Escombe, F. (1897-1898). Note on the influence of very low temperatures on the germinative power of seeds. *Roy. Soc. London Proc., Sect. B., 62,* 160-165.

Clark, D. C., Bass, L. N., and Sayers, R. L. (1963). Storage of hemp and kenaf seed. *Proc. Assoc. Off. Seed Anal. 53,* 210-214.

de Candolle, C., and Pictet, R. (1895). Sur la vie latent des graines. *Arch. des. Sci., Phys. et Nat. 33,* 497-512.

Dillman, A. C. (1930). Hygroscopic moisture of flax seed and wheat and its relation to combine harvesting. *Amer. Soc. Agron. J. 22,* 51-74.

Dillman, A. C., and Toole, E. H. (1937). Effect of age, condition, and temperature on the germination of flaxseed. *Amer. Soc. Agron. J. 29,* 23-29.

Eguchi, T., and Yamada, H. (1958). Studies on the effect of maturity on longevity in vegetable seeds. *Heratsuka Nat. Inst. Agr. Sci. Bul., Ser. E, Hort. 7,* 145-165 (in Japanese).

Esbo, H. (1954). Livskraftens bibehallande hos oskalat och skalat fro av timotj vid langtidslagring under ordinara betingelser pa fromagasin. *K. Landtbr. Akad. Tidskr. 93,* 123-148. (English summary, pp. 146-148).

Esbo, H. (1959). Livskraften hos timotejfrö under langtidslagring. *Lantbrhögsk. Inst. f. Växtodlingslara No. 12.*

Graber, L. F. (1922). Scarification as it affects longevity of alfalfa seed. *Amer. Soc. Agron. J. 14,* 298-302.

Griffith, W. L., and Harrison, C. M. (1954). Maturity and curing temperatures and their influence on germination of reed canarygrass seed. *Agron. J. 46,* 163-167.

Haferkamp, M. E., Smith, L., and Nilan, R. A. (1953). Studies on aged seeds. I. Relation of age of seed to germination and longevity. *Agron. J. 45,* 434-437.

Harrington, J. F. (1972). Seed storage and longevity. *In* "Seed Biology." Vol. 3 (T. T. Kozlowski, ed.), pp. 145-245. New York and London.

Harrington, J. F., and Thompson, R. C. (1952). Effect of variety and area of production on subsequent germination of lettuce seed at high temperatures. *Amer. Soc. Hort. Sci. Proc. 59,* 445-450.

Hermann, E. M., and Hermann, W. (1939). The effect of maturity at the time of harvést on certain responses of seed of crested wheatgrass, *Agropyron cristatum* (L.) Gaertn. *Amer. Soc. Agron. J. 31,* 876-885.

James, E., Bass, L. N., and Clark, D. C. (1967). Varietal differences in longevity of vegetable seeds and their response to various storage conditions. *Amer. Soc. Hort. Sci. Proc. 91,* 521-528.

Jensen, H. A., and Jørgensen, J. (1969). The influence of the degree of maturity and drying on the germinating capacity of *Festuca pratensis* Huds. *Acta Agr. Scand. 19,* 258-264.

Justice, O. L., and Bass, L. N. (1978). "Principles and practices of seed storage. Agricultural Handbook No. 506." Washington, D.C.

Koller, D. (1972). Environmental control of seed germination. *In* "Seed Biology," Vol. 2 (T. T. Kozlowski, ed.), pp. 1-101. New York and London.

Lakon, G. (1954). Der Keimwert der nackten karyopsen im saatgut von hafer und timothee. *Saatgutwirtschaft 6,* 259-262.

Lipman, C. B. (1936). Normal viability of seeds and bacterial spores after exposures to temperatures near the absolute zero. *Plant Physiol. 11,* 201-205.

Lipman, C. B., and Lewis, G. N. (1934). Tolerance of liquid-air temperatures by seeds of higher plants for sixty days. *Plant Physiol. 9,* 329-394.

MacKay, D. B., and Tonkin, J. H. B. (1967). Investigations in crop seed longevity. I. Analysis of long-term experiments with special reference to the influence of species, cultivar, provenance, and season. *Nat. Inst. Agr. Bot. J. 11,* 209-225.

McAlister, D. F. (1943). The effect of maturity on the viability and longevity of the seeds of western range and pasture grasses. *Amer. Soc. Agron. J. 35,* 442-453.

McNeal, X., and York, J. O. (1964). Conditioning and storing grain sorghum for seed. *Ark. Agr. Expt. Sta. Bul. 687.*

Mamicpic, N. G., and Caldwell, W. P. (1963). Effects of mechanical damage and moisture content upon viability of soybeans in sealed storage. *Proc. Assoc. Off. Seed Anal. 53,* 215-220.

Moore, R. P. (1972). Effects of mechanical injuries on viability. *In* "Viability of Seeds," (E. H. Roberts, ed.), pp. 94-113. London.

Moss, H. J., Derera, N. F., and Balaam, L. N. (1972). Effect of preharvest rain on germination in the ear and a-amylase activity of Australian wheat. *Aus. J. Agr. 23,* 769-777.

Oathout, C. H. (1928). The vitality of soybean seeds as affected by storage conditions and mechanical injury. *Amer. Soc. Agron. J. 20,* 837-855.

Ohga, I. (1923). On the longevity of seeds of *Nelumbo nucifera. Bot. Mag.* (Tokyo) *37,* 87-95.

Owen, E. G. (1956). The storage of seeds for maintenance of viability. *Commonwealth Agr. Bur. Pastures & Field Crops Bul. 43.*

Porsild, A. E., Harrington, C. R., and Mulligan, G. A. (1967). *Lupinus arcticus* Wats. grown from seeds of pleistocene age. *Science 158,* 113-114.

Roberts, E. H. (1972). "Viability of Seeds." London.

Riddell, J. A., and Gries, G. A. (1956). The influence of age and maturation temperature of wheat grains on plant development. *Ind. Acad. Sci. Proc. 66,* 62.

Sakai, A., and Noshiro, M. (1970). Some factors contributing to the survival of crop seeds cooled to the temperature of liquid nitrogen. *In* "Crop Genetic Resources for Today and Tomorrow" (O. H. Frankel and J. G. Hawkes, ed.), pp. 317-326. International Biological Program Publication 2 — London, New York, Melbourne.

Shands, H. L., Janisch, D. C., and Dickson, A. D. (1967). Germination response of barley following different harvesting conditions and storage treatments. *Crop Sci. 7,* 444-446.

Stevens, O. A. (1935). Germination studies on aged and injured seeds. *J. Agr. Res. 51,* 1093-1106.

Toole, E. H., and Toole, V. K. (1946). Relation of temperature and seed moisture to the viability of stored soybean seed. *U.S. Dept. Agr. Cir. 753.*

Toole, E. H., and Toole, V. K. (1954). Relation of storage conditions to germination and to abnormal seedlings of bean. *Internat. Seed Testing Assoc. Proc. 18,* 123-129.

Toole, E. H., Toole, V. K., and Gorman, E. A. (1948). Vegetable seed storage as affected by temperature and relative humidity. *U.S. Dept. Agr. Tech. Bul. 972.*

Weibull, G. (1952). The cold storage of vegetable seed and its significance for plant breeding and the seed trade. *Agr. Hort. Genet. 10,* 97-104.

Weibull, G. (1955). The cold storage of vegetable seed--further studies. *Agr. Hort. Genet. 13,* 121-142.

White, J. (1909). The ferments and latent life of resting seeds. *Roy. Soc. London Proc. 81 (B550),* 417-442.

Zeleny, L. (1954). Chemical, physical, and nutritive changes during storage. *In* "Storage of Cereal Grains and Their Products" (J. A. Anderson and A. W. Alcock, eds.), pp. 46-76. St. Paul, Minnesota.

GERMPLASM PRESERVATION: THE BASIS OF FUTURE FEAST OR FAMINE GENETIC RESOURCES OF MAIZE--AN EXAMPLE[1]

D. H. Timothy
M. M. Goodman

Departments of Crop Science and Statistics
North Carolina State University
Raleigh, North Carolina

In meeting the demands of increasing population and industrialization, society has reduced to an alarming degree the array of food production options. It has been estimated (Mangelsdorf, 1966) that man has used over 3000 species of plants for food and cultivated about 1500 species in sufficient quantity to have entered into commerce. Mangelsdorf states that about fifteen species actually feed the world. "These include five cereals: rice, wheat, corn, sorghum, and barley; two sugar plants: sugar cane and sugar beet; three 'root' crops: potato, sweet potato, and cassava; three legumes: the common bean, soybean, and peanut; and the two so-called tree crops: the coconut and banana."

More startling is our computation from world production figures (Food and Agriculture Organization of the United Nations, 1977) for wheat, rice, corn, barley, sorghum, oats, rye, the various millets, buckwheat, mixed grains, miscellaneous cereals, beans, peas, broad beans, lentils, chick peas, pigeon peas, cowpeas and all the other pulses, and the edible oilseeds such as soybeans, peanuts, sesame, rape, and sunflowers, which reveals that just three crops--wheat, rice, and corn--produce over 68% of the world's seed crop. Thus, the fate of millions hangs threadlike on the precarious balance of genetic systems of these three crops, their diseases and pests, and their interactions with environments.

[1]*Supported in part by NIH research grant number GM 11546 from the National Institute of General Medical Sciences, and by the Rockefeller Foundation grant in aid GA AGR 6905.*

ISBN 0-12-602050-7

The so-called Green Revolution is identified with two of these crops, rice and wheat. The developmental concepts and research efforts of the Green Revolution, slightly modified, are being used with the third crop, corn. That these efforts have increased food production is undeniable, but they have compounded the problems of increased genetic uniformity and the obliteration of genetic variability. The success of the new cultivars associated with the technologies and thrust of the Green Revolution is destroying the genetic variability that makes success of such programs possible (Chang *et al.*, 1972; Galinat, 1974; Harlan, 1972; Wade, 1972; Wilkes and Wilkes, 1972). Genetic resources are also being lost by increased grazing pressure, abandonment of old farming systems, and various developmental processes of a burgeoning population. These pressures have destroyed, and will continue to destroy, sources of as yet unknown but potentially valuable genes necessary for future plant improvement. The gradual loss of germplasm is usually referred to as genetic erosion, but the term genetic wipe-out (Harlan, 1972) is currently more appropriate and less euphemistic. The wipe-out is occurring not only in wheat, rice, and corn, but in hundreds of species. If it continues unabated, we place man's future in serious jeopardy. The most feasible recourse to lessen that hazard seems to be to assemble the germplasm resources of our cultivated plants and their relatives, and to preserve those genetic resources in germplasm banks.

Plant breeders and other researchers involved in varietal or hybrid development are not usually concerned with overall genetic diversity, but with lesser amounts of genetic variability or homozygosity for problems at hand. Duvick (1975, 1977) has suggested that to some extent the liabilities of a narrow gene base of the varieties grown in a given year within a region are sometimes partially offset over time by rapid development of new varieties; offsetting insurance is provided by maintaining older and/or less popular varieties, and by the regional variation among varieties in different zones of adaption. To the limited extent that different breeding organizations use different source materials, a certain amount of involuntary germplasm conservation is often practiced. At present, however, much breeding work is concerned with repeated backcrosses of a few outstanding performers to sources of single-gene attributes for currently desired incorporation. Furthermore, new materials are usually derived from crosses among adapted types, although a modest trend toward the use of adapted by exotic crosses or the use of widely based synthetics (perhaps even including some exotic germplasm) may be beginning (Duvick, 1977).

Traditionally, the rationale for germplasm preservation is usually based on chromosomal genes (Smith, 1971); for example, differences contained within the nucleus are due to additions, deletions, or substitutions of DNA segments in a chromosome, or differences are due to additions, deletions, or substitutions of multiple or partial chromosome sets. Unfortunately, the importance of cytoplasmic or extranuclear variation has been generally overlooked. However, the southern corn leaf blight epidemic of 1970 vividly illustrated the important differences among cytoplasms, as well as an agricultural vulnerability based on a single cytoplasm. Uniform cytoplasms are not only a potential hazard in plantings that utilize male sterility, but also in those widespread varieties derived from a common female background. The precise nature and location of cytoplasmic factors that control extranuclear inheritance of higher plants are not known, but differences in mitochondrial DNAs, and hence their genetic activity, have been correlated with disease reaction and male sterility/fertility in corn (Levings and Pring, 1976). Cytoplasmic inheritance is a useful phenomenon, and yet its occurrence in natural populations is poorly known and poorly understood. For years, phenotypic variation among cytoplasms has been amply demonstrated from genetic studies of male sterility and other manifestations of maternal inheritance. Only recently, however, has it been possible to attribute cytoplasmic variability to the DNAs of chloroplasts and mitochondria (Levings and Pring, 1976, 1977; Pring and Levings, 1978; Pring *et al.*, 1977).

GENE POOLS

The concept of a gene pool is a simple one. It is really nothing more than an assemblage of viable genetic variability. It is from such assemblages that man has been able to select genes that modify plants in a manner that man deems desirable. As with most simple concepts, the context in which it is used requires some definition, explanation, and restraint. In particular, the gene pool of the population biologist is often quite different from that of the plant breeder. The gene pool of the evolutionist or population biologist is usually the total genetic variation within a taxonomic unit, albeit a genus, species, or variety (Dobzhansky, 1951; Harlan and de Wet, 1971), whereas the gene pool of the plant breeder is usually much more limited in both scope and clarity of definition.

A gene pool, in most instances of current plant breeding and germplasm conservation usage, generally connotes the genetic variation of a specific population with at least a modicum of intermating, put together and maintained for some purpose other than maintaining the distinctive characteristics and genetic integrities of the indivdual components. Thus, gene pools, to many a corn breeder, are his private stocks of intermating populations to which additional material is often added. A gene pool is his source of new variation. It may be composed of locally adapted commercial types; it may be composed of local and exotic material; or it may be all exotic. To breeders of self-pollinated crops, a gene pool might be a large, bulked population composed of many different kinds of material. It may be a population composed of segregating descendents from any sort of hybridization, or it may have a built-in system of intermating by using male sterility.

A gene pool is also commonly identified in terms of its proposed use. It can be many things: the Stiff Stalk Synthetic of Corn-Belt maize; the genetic marker stocks of maize endosperm mutants--perhaps for high lysine content; the short, stiff-strawed Mexican or Japanese wheats; the wheat, soybean, or alfalfa varieties grown in this country 40 years ago; the extant collection of indigenous maize varieties of the Americas; the wild and weedy populations of wheat in the Middle East; a wild species of lintless cotton from a Mexican arroyo; a sample of pine trees from the Caribbean; seed from a small grain field in Ethiopia; a grove of fruit trees on the Crimean hillside; the cytoplasms in a collection of a wild Mexican grass. Each of these is an example of a gene pool.

Each gene pool contains a gene or group of genes that may or may not exist elsewhere. Perhaps of more importance are the arrangements of certain genes in individual chromosomes, their balance with those in other chromosomes, the stability of that whole chromosome structure, and the function of that structure in gene-cytoplasm interactions. The population structure of primitive cultivars or landraces and their related wild and weedy species is a highly integrated system of genetic and environmental balance. It is to these gene pools that man has continually turned in the search for genetic material to improve his foodstuffs. It is to these gene pools that he must turn for present and future plant breeding needs. For millennia, these populations, subjected to natural and artificial selection, have served as conservatories of the heredities of our plant resources.

CROPS, WILD SPECIES, AND WEEDS

Somewhere in the haze of antiquity man began to select and domesticate plants, apparently as a series of independent happenings in widely separated areas (Baker, 1971; Harlan, 1975a). Archeological and botanical evidence indicates great amounts of initial variability during the first stages of domestication followed by increasing phenotypic uniformity as the crop became increasingly domesticated (Mangelsdorf, *et al.*, 1967; Oka and Morishima, 1971). Periodic infusions of new germplasm from wild or related species and cultivars released new genetic combinations, sometimes in an explosive display of diversity. Selection among the myriad offspring of this array often resulted in enormous jumps of productivity and, within any given locale, a gradual return to visual uniformity.

Man in his travels and migrations carried his foods from one region to another. Throughout this process the plant was continually subjected to the rigors of each environment into which it had been thrust. On top of such "natural" selection, man also imposed his own selection criteria for certain characteristics. The interaction of man, the plant, and the environment went on for hundreds and hundreds of years. Thus, for any one species, certain varieties developed in one region, while other similar but distinct varieties developed in other regions (Hussaini *et al.*, 1977). These varieties are called primitive cultivars, indigenous varieties, races, farmer varieties, or landraces. The characteristics of these were and are as varied as their uses, the people who grew them, and the environments in which they were grown. The enormous stores of genetic variability contained in the landraces are now being lost at a continually increasing rate.

Wild species closely related to cultivated plants are important as occasional and natural genetic contributors to our crops, and also as possible progenitors of the economic species. Our understanding of how and from what our food plants originated is vague in most instances, but is becoming more clearly understood in others. Interest in the origin of crops is more than academic. In some cases, the immediate predecessor from which the cultivated species evolved is extinct, but a primeval source may still be extant. Some crops may have evolved as polyploid derivatives of one or more species, others may be simple diploids, and some may have arisen from weeds of a different crop.

Origins are sometimes complex, but once understood, illustrate an evolutionary road map. By following those same pathways, artificial hybridization with an ancestral wild species can often be used more easily to bring a desired gene into the cultivated relative than would be the case using a more distant relative. The substituted chromosome carrying the gene from the wild relative will have additional genes that are undesirable in an economic plant. Breaking that linkage and recombining the alien gene into the appropriate chromosome of the economic species is easier if the chromosomes are reasonably homologous. Other things being equal, a putative parental species would be used for gene transfer rather than a more distantly related relative. However, the use of wild relatives is not restricted to those of closest relationship. Closely and distantly related species have been used in a number of ways to transfer desirable characters to wheat, cotton, tobacco, and rice (Beasley, 1942; Chang *et al.*, 1972; Gerstel, 1945; Sears, 1956).

The weedy relatives associated with landraces are at various stages of intermediacy between the cultivar and its wild relative. They may accommodate gene exchange in either direction. The variation of a weedy population can be enormous. One segment of the population may mimic the landrace at a particular growth stage, another segment may flower at the same time as the cultivar, yet another may be easily spotted as closely resembling the wild plant (Wilkes, 1972a). Continued association, gene exchange, and selection of cultivars and their weedy relatives have resulted in weedy populations that assume racial properties (Chang, 1976; Wilkes, 1977). They are highly integrated genetically and buffered cytologically. The blocks of genes from wild and cultivated parents have been broken down through long periods of time by various recombinations of the wild and cultivated genes.

Sterility of hybrids between weedy and cultivated forms is not as severe as that often encountered in crossing wild and cultivated species (Harlan *et al.*, 1973). The value of weedy relatives has often been grossly overlooked and their directed exploitation for improving cultivated plants is practically nonexistent. Most often collectors tend to bypass them for cultivated varieties and wild species. Even worse, most plans for the collection and preservation of germplasm do not include the weedy forms.

The populations of landraces and their wild and weedy relatives are genetically balanced with the environment. This allows the frequencies of various portions of the populations to ebb and flow in response to natural selection pressures. The greater the diversity of the population, the more plastic its response may be.

The basic population structure is determined principally by the mode of reproduction (Stebbins, 1950). Individuals in the cross-pollinators are generally in a highly heterozygous state, and each plant of the population is essentially distinct from all others. The self-pollinators are composed of great numbers of homozygous individuals. However, the self-pollinators have appreciably more diversity, and especially more heterozygosity, than is often realized (Allard, 1965; Allard *et al.*, 1968; Marshall and Allard, 1970). There is usually a small percentage of out-crossing, and although the return to homozygosity is rapid, occasional out-crossing permits a continuous source of new recombinations, first as segregating heterozygotes and then as a series of stabilized homozygotes.

Some species--cotton and sorghum, for example--are intermediate in their mode of reproduction, and have characteristics common to both selfing and out-crossing species.

Asexual species are of a special nature, but they may contain tremendous stores of variability. Their variability is released by occasional out-crossing to nearby related and sexual species, or by a rare breakdown of asexual control and the subsequent completion of fertilization (de Wet and Harlan, 1970; Harlan *et al.*, 1964). In addition to its relationship to population structure, the mode of reproduction affects collecting or sampling techniques and maintenance of germplasm (Frankel and Bennett, 1970; Frankel and Hawkes, 1975).

PREVIOUS COLLECTIONS: GOOD AND BAD

With a few exceptions, our past efforts at plant collection have been extremely pragmatic. The collections were sporadic, unsystematic, poorly funded, and usually the effort of a few men racing against time. Too often, we responded as a reaction to a specific need, frequently a disease- or pest-related problem. An expedition would be organized, collections made and distributed to researchers who would, it was hoped, isolate the resistant gene to be incorporated into breeding material, and then the collection would be discarded. When the next crisis occurred, we went through the same orchestrations. The wild and landrace populations were considered as everlasting founts. The concept was, of course, erroneous. The complete disappearance of wild populations and landraces from areas of thousands of square miles is being documented repeatedly (Frankel and Bennett, 1970; Committee on Genetic Vulnerability of Major Crops, 1972). Much of the variability from those areas has been lost. In some cases it all would have been lost had it not been for all-too-few farsighted efforts to collect some of that germplasm. An excellent, unfortunately rare example is the case of the corn varieties that preceded the famous Corn Belt hybrids.

Fortunately, we do have many thousands of individual collections of our major crops. For example, the world collections may contain 26,000-30,000 wheats (Harlan, 1972), 22,000 sorghums (Webster, 1976), 12,000-14,000 rices (Chang *et al.*, 1975), and 1000 or so finger millets. The USDA wheat collection contains over 19,000 accessions, mostly assembled since 1948 because the original collections were lost (Committee on Genetic Vulnerability of Major Crops, 1972). The U. S. peanut collection contains approximately 6000 entries (Hammons, 1976), whereas cotton and soybeans number around 3000 (J. A. Lee, personal communication) and 7000 (C. A. Brim, personal communication), respectively. Maize accessions in Latin American germplasm banks approximate 24,000 (Brown, 1975).

Individual researchers are the principal agents in maintaining significant portions of these collections. Most of the wild and related species are maintained, as are the cytoplasmic, chromosomal, and genetic marker stocks, at the individual discretion, effort, and initiative of a handful of persons. None of the states has a suitable arrangement for the maintenance of germplasm on a broad scale. The National Seed Storage Laboratory at Fort Collins is not expected to fill the need for some time to come, if ever (Current Policy Statement, National Seed Storage Laboratory, U. S. Department of Agriculture, mimeographed, undated). Recent reorganization of the USDA, aside from its benefits, has demolished national leadership pertaining to specific crop plants. The recently organized National Plant Germplasm System was designed to meet the highly variable needs of the U.S. plant scientists and will provide a state-federal forum for considering matters of mutual importance, apparently with appreciable consideration given to both U.S. regional and national needs (Agricultural Research Service, 1977). It remains to be seen if the new organization will permit the high level of performance needed. Although those associated with the former Plant Introduction Service and the New Crops Research Branch have done their utmost (Burgess, 1971), they have been severely limited by several inadequacies. Most of the forage collections, for example, were and are maintained at various USDA Plant Introduction stations. Many of these collections are cross-pollinated. Yet the Plant Introduction Service was forced to plant these collections in short rows, one collection beside another, and allow fertilization to occur without pollination control. A collection originating from Turkey may then be fertilized by others from Greece, Spain, Algeria, or France. The researcher who subsequently tries to use seed increased or rejuvenated from that Turkish collection will have no idea of what kind of material he is really working with. If he is trying to locate geographical sources of certain genetic characters, he is defeated, or severely handicapped, before he starts. This is not a condemnation of the Plant Introduction personnel; they are dedicated people. They do the best they can with inadequate funding, inadequate facilities, and inadequate numbers of professional and subprofessional personnel. It is the same story in most areas of the world.

Appreciable portions of these collections are redundant. Individual accessions within them may have gone around the world several times with intervening stops at various experimental stations. Each time they reenter this country they are given a new accession number. Many were collected over periods of years from the same area, sometimes at the same site. The collection areas were limited to those politically accessible, and oftentimes along roads that were the shortest distance between cities. Now, there is nothing wrong with this, but it does create a false sense of security when collections are considered in terms of numbers only. As a consequence of the collecting and accessioning procedures, our samples of germplasm are genetically much narrower than their numbers would indicate at first glance. Even at second and third glances, the documentation and peregrinations of many of the collections preclude tracing them to their geographical origin. In addition to the redundancies of the collections, there are geographical and evolutionary voids.

Systematic collections are needed. The variation of each crop species and its relatives must be sampled in areas of differing ecology and culture, especially in zones of great variation. It matters little if these areas are referred to as Vavilovian centers of origin, areas of diffuse origin, centers of diversity, microcenters, or whatever. What matters is that the variability within those areas be maintained. However, there are problems in doing that. An illustration of the situation with corn might be helpful. Under ordinary circumstances the details presented would only be of interest to specialists. These details, however, are often unavailable even in the reports of the various germplasm committees, and the maintenance of germplasm is critically dependent upon such details. The maize collections have probably been studied, documented, described, and maintained more thoroughly than those for any other crop. Thus germplasm resources for other crops are likely to be even less well preserved than those for maize.

MAIZE AND ITS RELATIVES

There is good evidence that at least one kind of cultivated corn originated in the Tehuacan Valley of Mexico (Galinat, 1971, 1977; Mangelsdorf *et al.*, 1964, 1967; Mangelsdorf, 1974). To date, there is no other archeobotanical evidence indicating a different site of origin. For several thousand years, the small cobs of this primitive plant, now extinct, sustained its cultivators. Then the primitive corn began to accumulate new characteristics, presumably from the incorporation of germplasm from a related wild grass, teosinte, and to assume the proportions of present day landraces.

Today in the Valley of Mexico, teosinte, the same species that contributed to maize evolution thousands of years ago, is found as a weed in maize fields. Its reproductive isolation from maize is not complete, and so a small percentage of hybrids and backcrosses are generated each season. These fields are microcenters of evolutionary activity between a crop and its weedy relative. Other forms of teosinte are found in different regions as a wild plant not intimately associated with the cultivated crop (Galinat, 1972; Wilkes, 1967). Teosinte has distinct races, and its distribution is limited to Meso-America. All described races are annual. The tetraploid perennial species is extinct, except for individual plants grown in greenhouses and experimental gardens.

Tripsacum, the other relative of maize, is perennial and can be crossed experimentally with maize but not with teosinte. Its role in the evolution of maize is not clear, although it may be associated with certain characteristics of South American maize. The 11 described species of tripsacum form a polyploid series found in certain habitats from Connecticut to Paraguay (Cutler and Anderson, 1941; de Wet *et al.*, 1976; Hernandez and Randolph, 1950; Randolph, 1970).

Only in the last 15 years has a reasonable collection of the maize relatives been attempted. A fairly systematic collection of teosinte is now in hand (Wilkes, 1972b), but the tripsacum collection is probably less than a thousand plants. These collections were made by students of maize with sporadic funds from philanthropic or granting agencies and have been maintained under all manner of cooperative word-of-mouth agreements among the interested scientists. Only during the last 5 years has an institutional interest with suitable facilities and long-range probabilities been indicated.

MAIZE COLLECTIONS, RACES, AND MAINTENANCE

When Wellhausen and his colleagues initiated the cooperative corn program of the Mexican government and the Rockefeller Foundation, they began by collecting the local varieties of corn. These indigenous strains were to be the basis of the breeding program. The collections soon became a hodgepodge of incomprehensible variability. To bring order out of chaos, the indigenous strains were classified into races and the study was published as the "Races of Maize in Mexico" (Wellhausen *et al.*, 1952). This classical example was followed by a series describing the races of maize in South America, the Caribbean, and Central America (Brieger *et al.*, 1958; Brown, 1960; Grant *et al.*, 1963; Grobman *et al.*, 1961; Hatheway, 1957; Ramírez *et al.*, 1960; Roberts *et al.*, 1957; Timothy *et al.*, 1961, 1963; Wellhausen *et al.*, 1957). From some 11,000 collections of indigenous varieties in the Western Hemisphere, over 280 races of maize were described. Morphological, physiological, genetical, and geographical characteristics were included, and in some cases cytological and ethnobotanical information as well. It was intended that the racial descriptions be preliminary and that they serve as a logical starting point for additional studies of maize, its evolution, and its utilization. Studies of this nature have been limited (Bird and Goodman, 1977; Goodman and Bird, 1977; Hernandez and Alanis, 1970); a general survey of the races of maize by Brown and Goodman (1977) provides an overall view of the races of the Americas; the present-day races of maize are postulated by Mangelsdorf (1974) to have descended in six lineages from wild races.

The maize collections were initiated at three primary germplasm banks: Chapingo, Mexico; Medellín, Colombia; and Piracicaba, Brazil. While procedures and results differed somewhat among these centers, all were faced with common problems. A brief description of some of the procedures, goals, and achievements of the Andean center at Medellín suggests the scope of the project.

The Andean collections were usually made by obtaining 10 to 15 ears of each sample from farmers' fields and houses, granaries, and marketplaces. The ears were sent to Medellín, Colombia, where they were cataloged, documented, and measured for numerous characteristics. The ears were shelled, except that three ears of each sample were retained as museum specimens; only two rows of grain were removed from the specimen ears. These large samples of seed were put into storage for maintanance. As precaution against loss, smaller duplicate seed samples were put in cold storage at Medellín and also sent to the seed storage center at Glenn Dale, Maryland, maintained by the Division of Foreign Plant Introduction of the USDA.

The inadequacies of the Medellín storage facilities were alleviated over a 10 year period. In the interim, it was necessary to rejuvenate seed stocks periodically to maintain germination. Corn is cross-pollinated and heterozygous; the integrity of the collection could be lost very quickly by natural selection, a too small number of plants involved in the increase, or improper pollinating procedures. Attempts to prevent genetic loss were made by using careful pollinating techniques in populations as large as possible, using open pollination in large blocks spatially isolated from other maize, planting single collections on three separate dates to allow for differences in flowering, planting at appropriate altitudes, or sending the long-day responsive stocks to Mexico or Iowa. There were some genetic shifts and losses of complete samples, but by and large the effort was successful, although extremely costly in terms of manpower, money, and the use of experiment station facilities. It was apparent that the operation of the cooperative Colombian Government-Rockefeller Foundation corn improvement program and the maintenance of the individual samples of the Andean Maize Germplasm Bank could not continue indefinitely at the same level of operation.

After the indigenous strains of maize from each of the Andean countries had been classified into races, the decision was made to begin forming racial composites of the individual strains according to that biological classification. It was also decided that certain of the individual collections of each race should be maintained individually. The procedure was initiated with the collections comprising the races of maize in Colombia. From among all the collections from that country designated as representative of a race, usually three to five strains were chosen as "type" or "typical" examples of that race, and these were individually maintained and increased. Taxonomically, these would be analogous to syntypes. The other equally representative collections of that race were designated as "others"--taxonomically analogous to paratypes. Subsequent monographs that described the Andean races listed the collections as "types" (Bolivia, Chile, Ecuador) or "typical" (Venezuela, Peru) and "others." The compositing system consisted of mixing together equal numbers of viable seeds (as determined by germination tests) from each collection. Race "A" composite therefore included the "type" or "typical" collections as well as "others," that is, the composite was made from syntypes and paratypes. Additional composites from each race were sometimes made, for example, of only the "type" collections. "Some races contained subgroups differing, for example, in grain color or kernel characteristics. Therefore, if race 'C' had both yellow and white grain [subgroups], and also flour and flint starch texture [subgroups], there may have been five different composites made for this race [within each of the collection groupings of "type" and "type" plus "others"]: White flint, yellow flint, white flour, yellow flour, segregating for starch and color. Likewise, the collections intermediate between races 'A' and 'B' were [often] composited to form one population of 'A-B' germplasm (Timothy, 1972, p. 649). In this paper, the individually maintained "type" or "typical" collections will be referred to as "type" collections, and composites of syntypes and/or paratypes will be referred to as typical composites.

This system of germplasm preservation is a compromise, but it maintains a few "type" individual collections of each race and still permits maintenance of large seed supplies of each of the typical racial composites. "Numerous requests from all parts of the world are more easily filled. It also allows more thorough study and evaluation of native races to determine the sources of genes for yield, insect resistance, and other economic characteristics" (Timothy, 1972, p. 649). A request was made for approximately 5 kg of each typical racial composite and each "type" collection from this increase to be sent to the United States for long-term storage at the National Seed Laboratory at Fort Collins, Colorado. Compliance with that request was begun. Large quantities of additional seed of that same material were stored at the Colombian Germplasm Bank.

By 1963, descriptions of all the known races of maize in Latin America had been published. Much of the North American corn was preserved and described. Workers in Africa, Asia, and Europe were collecting, preserving, and cataloging the races there. For the first time, the variation of an important world crop would be categorized in units of workable size representing easily recognized groups. Moreover, the variability of that crop would be preserved as a legacy for the future. At least that was the thought.

CURRENT STATUS: THE LATIN AMERICAN MAIZE COLLECTIONS

Once germplasm is collected, cataloged, described, and stored away in a freezer it tends to be forgotten. The attitude seems to be that it was job well done and now we must get on with other things. Also, over relatively short periods of time there are changes in personnel and institutional attidudes and policies. Additionally, there is usually little funding and little or no program for maintaining and studying germplasm collections. In 1963, however, the general impression was that the corn germplasm collection was in pretty good shape.

By 1968, it was apparent that despite the extensive preparations that had been made to preserve maize germplasm, problems had begun to arise. (Many of the details and arrangements in this and subsequent paragraphs are based on personal knowledge and experience). The germplasm banks in Brazil, Colombia, Mexico, and Peru, which maintained all American collections except those from the United States and Canada, were all faced with maintenance problems similar to those referred to above. There were other problems as well: numerous breakdowns of refrigeration equipment, power failures, various strikes or civil disorders which prevented personnel from entering the facilities, and drought or flooding during the growing seasons when seed increases were made.

Duplication of material and effort was no guarantee of preservation. For example, the Chilean collections were sent to Mexico and Iowa for seed increase and for recording of plant data to be used in describing the Chilean races. The data and seed from Mexico were sent back to Colombia by air shipment. Both were lost. The duplicate data books and seed samples retained in Mexico for such an eventuality were also lost in a flood (the agronomist responsible for them died in the disaster). In short order, rejuvenated seed of the collections from the Chilean highlands and the data from those collections were wiped out, although the lowland Chilean increases and data recording by Pioneer Hi-Bred International in Iowa were successful.

Most all of the standby collections--duplicate samples of about 4 ounces or 200 seeds of each of the original collections--had been shipped to Glenn Dale, the USDA's Maryland Plant Introduction Station. These numbered about 11,000 entries (Committee on Preservation of Indigenous Strains of Maize, 1954, 1955). From there they were sent in the mid-1960s to the National Seed Storage Laboratory, USDA-ARS, Fort Collins, Colorado, but they were not officially accepted because of variable germination, small seed lots, lack of an agreement for rejuvenation of viability, and perhaps other reasons. After negotiation with the International Maize and Wheat Improvement Center (CIMMYT), Mexico City, the Cuban, Guatemalan, and South American collections were shipped from Fort Collins to CIMMYT (about 7600 entries), and the remainder were discarded by Fort Collins. Most (about 600) of the Bolivian collections were grown out in the winter of 1969 at Tepalcingo by CIMMYT. Only a few produced seed; the remainder were lost. Thus, 7 years after the last of the race bulletins appeared, the standby collections were reduced from about 11,000 entries to 7000 entries, and the germplasm collections of two countries (Mexico and Bolivia) and one region (the West Indies) were essentially eliminated from the group.

The status of the individual collections at the various germplasm banks varies greatly. Some collections are classified by race, some are not. Some of those classified are listed in the race bulletins or elsewhere as being "type" collections; most are not. Only the status of the "type" collections is reasonably known at present, although a cataloging process of many of these collections is under way (Information Sciences/Genetic Resources Program, 1977). It is suspected that the status of the other collections is poorer, but this may not always be so. CIMMYT until recently had only relatively few individual "type" collections from Mexico, Central America, and the Caribbean, despite a rather large number of accessions. Many of CIMMYT's individual collections have been assembled in recent years by E. Hernandez X., A. Blumenschein,, José Jiménez, and Pablo E. Daza B. as the bank was started only in 1960, after the original collections had been completed. Many others were deposited at CIMMYT by the Brazilian bank at Piracicaba, which now maintains only racial composites (personal communication, M. Gutiérrez., CIMMYT; Kashiwakura and Paterniani, 1972). Consequently, seed requests for collections that were used and documented in describing "The Races of Maize in Mexico"

(Wellhausen *et al.*, 1952) have often been filled with seed supplies from other collections.

The Instituto Nacional de Investigaciones Agricolas (INIA) has the most complete set of Mexican, Guatemalan, and Caribbean individual collections. INIA inherited the germplasm bank of the cooperative program of the Rockefeller Foundation and the Mexican government. (Some Guatemalan collections have been salvaged by CIMMYT from the standby collections formerly at Fort Collins, but many Guatemalan collections are quite difffi-cult to maintain at CIMMYT). Many of INIA's Guatemalan collec-tions are original (nonincreased) seed. Neither CIMMYT nor INIA has a frost-free high altitude experiment station that would enable them to maintain late maturing, high altitude collections.

Most of the individual collections from eastern South America (the Guianas, Brazil, lowland Bolivia, Paraguay, Uruguay, and Argentina) made by the Institute of Genetics, Escola Superior de Agricultura "Luiz de Queiroz," Universidade de São Paulo, at Piracicaba in Brazil, are no longer maintained there. That bank did not utilize modern cold storage equipment and has been essentially phased out (Kashiwakura and Paterniani, 1972). It is to be replaced by a new bank under the auspices of EMBRAPA (Empresa Brasilera de Pesquisa Agropecuaria). Most of those individual collections formerly stored at Piracicaba have been increased by CIMMYT either from the standby collec-tions from Fort Collins or other sources, however, and are still available.

Many of the Peruvian collections are maintained by the Programa Cooperativa de Investigaciones en Maíz, Universidad Agraria--La Molina, Lima, Peru, which assumed responsibility for them, leaving the Colombian bank with responsibility for the other Andean collections. Many of the Peruvian collections adapted to altitudes of about 2000 to 2800 m were lost due to lack of facilities to increase collections at those altitudes. Efforts are under way in Peru to recollect and replace the representative materials which have been lost. In addition, the Peruvian portion of those standby collections from Fort Collins that are no longer available elsewhere is being increas-ed in Peru under an agreement with CIMMYT.

The collections from Venezuela, Colombia, Ecuador, Peru, Bolivia, and Chile were originally to be stored in Colombia at what is now the Instituto Colombiano Agropecuario (ICA) with standby samples at Fort Collins. Almost all the standby samples made it to Fort Collins (and then to CIMMYT), but the Chilean and Peruvian samples are not currently available from Colombia.

A portion of the story of the Chilean collections has al-ready been presented. Fortunately, the low altitude Chilean "type" collections were increased in the United States by

W. L. Brown, of Pioneer Hi-Bred International, and placed in Fort Collins. They are still there and appear to have good germination. Those collections remained at Fort Collins even though the standby collections were sent to CIMMYT. In addition, it has been possible to salvage a number of the high altitude Chilean "type" collections from the well-traveled standby collections (now at CIMMYT).

The Peruvian collections unfortunately were given two sets of collection numbers. One set was used in Peru and published (Grobman *et al.*, 1961), while another set was used in Colombia, Fort Collins, and in the reports by the Committee on Preservation of Indigenous Strains of Maize (1954, 1955). Thus, use of the Peruvian seeds from the standby collections from Fort Collins to replace lost collections in Peru has been hindered by the lack of a complete cross-listing of the two sets of collections numbers. No cross-listing of the two sets has apparently ever been published. Adequate safeguards for the Peruvian collection were further hampered because the complete set of Peruvian collections was never received in Colombia; hence no complete set of standby collections was ever assembled.

The Bolivian "type" collections stored in Colombia were neglected for several years but are now being increased for tentative transfer to CIMMYT (Brown, 1975). It appears that a number of them (mostly high altitude materials) have been lost, but until the increases have been completed, their exact status must remain in doubt. Duplicate samples of many of the low altitude Bolivian "type" collections had been saved by W. L. Brown of Pioneer Hi-Bred International. These have been increased in Florida and sent to CIMMYT as a precaution against still further erosion of the Bolivian collections. Unfortunately, Mario Gutiérrez, who rescued CIMMYT's maize germplasm bank from chaos in the mid-1960s, and who was responsible for salvaging not only many of the well-traveled standby collections but also most of the collections from the former bank at Piracicaba, Brazil, is no longer at CIMMYT. As a result, many of the plans (such as those for the transfer of the Bolivian collections to CIMMYT) for the preservation of maize germplasm at CIMMYT seem unlikely to be achieved.

The Colombian bank will remain in charge of the collections from Venezuela, Colombia, and Ecuador. The "type" collections from Venezuela and Colombia are generally viable and available from ICA. In the recent past, at least, this has been much less so for Ecuador, especially highland Ecuador, but the standby collections (now at CIMMYT) have filled the gap reasonably well. Many of the Colombian "type" collections were increased and placed in Fort Collins (the only country for which this seems to be true). These increases remained in Fort Collins when the standby collections were removed.

In addition to individual collections, there are several kinds of composites, most of which remain poorly described, if at all. The typical composites of most Colombian races are still at Fort Collins. The typical Colombian composites, as well as the typical composites of most races from Venezuela and Ecuador, are available from ICA. A smaller proportion of typical composites from Bolivia is available from the same source. The latter three sets (Venezuela, Ecuador, and Bolivia) apparently were not deposited in Fort Collins. Composites were never made for the Chilean races, but composites for most of the Peruvian races were made at La Molina, Peru, where they are still generally available. The composites developed at the Brazilian germplasm bank were not deposited in Fort Collins but were sent to Mexico and, for the most part, are available from CIMMYT (Paterniani and Goodman, 1978). (No "type" collections were ever designated for most of the races of eastern South America). Mexican racial composites of uncertain origin are still at Fort Collins. Outside the Andean region, and perhaps within that region in recent years, some collections may have been assigned to a race even though they were not typically representative of that race. Perhaps they possessed more characteristics of that race than of any other, but admixtures from other sources should have precluded their inclusion in representative racial composites or as representative specimens of a particular race.

Apparently there was some indiscriminate compositing of the individual collections in certain maintenance programs, without proper assurance that even the individual "type" collections were also being maintained. From the plant breeding viewpoint, this can be an acceptable and very logical procedure. In fact, the formation of complex composites of *unrelated* materials is often indicated, insofar as the immediate needs of feeding people by modern agricultural production is concerned. Broad-base composites are often thought of as gene pools, and a breeding program frequently has several of them. But as a procedure for preserving genetic resources they are completely unacceptable as they result in the loss of the biosystematic identity and the genetic integrity of their individual components.

In Brazil and during certain periods at CIMMYT, but not at INIA, in Mexico, the development of racial or subracial composites apparently took precedence over the maintenance of many "type" or other individual collections. In fact, until recently, relatively few "type" collections from Mexico and Guatemala were stored at CIMMYT. Before CIMMYT was organized most of these collections had been stored at INIA; relatively few of them, usually collections of special interest, were utilized in the corn breeding programs at CIMMYT. As a result of the emphasis on composites, some of the individual and "type" collections were lost. In the case of the Brazilian individual collections, the duplicate or standby samples (now at CIMMYT) were available to replace many of the lost collections. However, the Mexican samples among the Fort Collins standbys were discarded before it was realized that many of them were not available in Mexico.

Tables I and II summarize the general status of the "type" collections of Latin American maize, when such collections have been designated, and the status of the racially classified collections when "type" collections were not designated. A number of the high altitude collections from Ecuador are in the process of being salvaged from the standby Fort Collins collections, so their availability is limited. The status of collections from Bolivia and Chile is critical, and much of the maize of Central America (except Guatemala) remains undescribed.

TABLE I. Status of the Number of Individual "Type" Collections of the Latin American Races of Maize as of April 1978

Country or region	*Typical collections listed*	*Typical collections still available*
Mexico[a]	*154*	*132*
Guatamala	*180*	*115*
Honduras[b]	*1*	*0*
El Salvador[b]	*2*	*2*
Nicaragua[b]	*1*	*0*
Costa Rica[b]	*3*	*2*
Cuba[c] *and the West Indies*	*49*	*39*
Venezuela	*86*	*86*
Colombia	*129*	*127*
Ecuador	*154*	*150*
Peru	*183*	*154*
Bolivia[d]	*141*	*106*
Chile[e]	*80*	*56*
Argentina[c,d]	*26*	*23*
Paraguay[c,d]	*13*	*9*
Totals	*1202*	*1001*

[a]*Includes the collections of Hernandez and Alanis (1970).*
[b]*Essentially unstudied.*
[c]*Fairly well collected and studied, but few if any "type" collections have been documented.*
[d]*See Table II also.*
[e]*Many of these in immediate danger of complete loss.*

TABLE II. Current Status of Racially Classified Individual Original Collections from the Brazilian Germplasm Bank[a]

Country or region	*Collections classified*	*Collections still available*
Argentina	*57*	*50*
Uruguay	*81*	*20*
Paraguay[b]	*48*	*36*
Brazil[b]	*984*	*577*
Guianas[b]	*21*	*12*
Bolivia	*19*	*18*
Totals	*1210*	*713*

[a]*No "type" collections have been identitied for most of the races described at that bank (does not include any post 1965 collections, which have yet to be described and documented).*

[b]*The western part of Paraguay, much of Amazonas, and the less accessible parts of the Guianas are still largely uncollected.*

Several described races have apparently been completely lost (Polulo and Negrito from Chile, several subraces of Quicheño from Guatemala, Amarillo de Ocho and several subraces of Capia from Argentina, Harinoso Dentado from Colombia). All the "type" collections have been lost for the races Rienda and Jora from Peru, Paru from Bolivia, and Coastal Tropical Flint from Dominica. In addition, W. H. Hatheway's collections from Cuba apparently were not deposited in any of the germplasm banks.

Finally, the North Central Plant Introduction Station of the USDA at Ames, Iowa, has a large collection of U. S. Indian corns assembled principally by W. L. Brown, A. E. Longley, and H. C. Cutler (many of these were salvaged by Longley, later increased by Brown, and redeposited with the USDA after the USDA had discarded them), as well as an abundance of poorly documented open-pollinated varieties and miscellaneous undocumented plant introductions (from various catalogs and seed lists of the North Central Regional Plant Introduction Station, Ames, Iowa). These have never been studied in the same detail as have the Latin American races. The USDA itself has never had a comprehensive collection of Latin American maize, but the materials at Ames are generally believed to be well maintained.

That, in brief, is the status of the collections of American maize. What is going on now is a last-ditch stand (Brown, 1975) to prevent further loss of something which many people had assumed was well preserved. The situation with other crops is probably not any better. Indeed, it appears uniformly worse.

CONCLUSIONS

Germplasm resources are fast disappearing, and there is urgent need to collect and preserve those resources. In the past there have been mistakes of omission and commission, inadequate support in maintaining the integrity of germplasms, unfavorable weather conditions, and so forth. But perhaps most damaging to the maintenance of germplasm and the integrity of its individual components is the concept of manipulating the formation of gene pools. If the concept of a gene pool, insofar as germplasm maintenance is concerned, could be likened to a military motor pool composed of separately usable units, rather than a beachcomber's stew pot into which everything was dumped and blended, we would be much better off.

To use genetic variability most intelligently, we must know where, and preferably how, it originated, not only for disease or insect resistance but also for yield and quality features of the market. As we search for these genes, it is increasingly clear that if we understand the evolutionary relationships of the crops and their relatives, modification of the crops to suit our needs becomes easier. To do all this requires that the essential integrity of the germplasm be preserved. Only when all individual collections cannot be maintained should composites be initiated. Compositing should be done only on a biologically systematic basis with as many categorical units as possible. (An excellent example of where it is much better to be a splitter rather than a lumper). We do not deny that progress, even occasionally spectacular progress, can be made on a hit-or-miss basis where the breeder knows virtually nothing about the sources of his material. However, long-term progress depends upon accumulated knowledge of source materials and guaranteed access to them.

Methods of conserving genetic resources vary according to the crop. Each nation cannot maintain a complete germplasm collection for each crop. The scope and cost of the program would be too large and much material would be unadapted.

Various national, philanthropic, and multinational entities have increased their interest in germplasm resources (Harlan, 1975b), but much of the activity has been of a survey-discussion nature and often reiterative, with one group repeating the work of another albeit at different levels of planning and/or organization. A few action programs have been initiated within the past ten years, but private opinions differ from official positions in regard to success. However, awareness of the germplasm problem is reaching higher levels of institutional management, and it is hoped that well-founded programs will emerge from the scores of committee reports, working papers, and organizational charts. The ponderous natures of governments and multinational organizations do not seem to offer much immediate hope for solving the problem. Several of the philanthropic foundations and some of the international research centers that they support (CIMMYT, the International Rice Research Institute, and the like) are hampered less by the demands of protocol and seem to offer the best possibility for immediate and interim germplasm maintenance facilities until additional organizations are properly established and operational. Current efforts of the IBPGR (International Board for Plant Genetic Resources, 1975, 1976) to collect germplasm from critical areas are bearing fruit, although the ultimate success of those efforts in germplasm preservation remains to be seen.

"The maintenance of a living collection is usually regarded as a routine and time-consuming, yet essential, task. Much of the material maintained there seems to be of little current interest--occasionally, of course, a threatened epidemic or new insight into a disease problem will generate a sporadic interest in screening everything available. A great deal of the material that leaves the bank is discarded; it is regarded as a gift, not as a loan. Surely a more suitable arrangement could be devised. One possibility is the development of germplasm maintenance centers in which maintenance is regarded as the primary goal, not as a by-product of breeding activities. In this connection it is worth noting that the plant breeder, entomologist, or plant pathologist tends to regard a germplasm collection as an inexhaustible source from which he can extract only such experimental material as is of interest for his specific purposes. On the contrary, the experimental taxonomist, crop plant evolutionist, and ethnobotanist consider the entire collection to be necessary experimental material--quite apart from its potential economic value. They need to preserve and study as wide a variety of material as they can manage and to use it continuously. Because all classifications have to be revised as new material is collected and new analytical procedures developed, the experimental taxonomist's work is never done. A germplasm center could effectively serve a dual purpose as experimental material for several disciplines and as a reliable and continuing source of germplasm for the plant breeder. Coupling these two objectives could relieve the plant breeder of a routine chore and provide the taxonomist and others with facilities and experimental material not now available to them" (Committee on Genetic Vulnerability of Major Crops, 1972, p. 301-302). It is clear from the example with maize that germplasm banks have not functioned satisfactorily, either as germplasm banks per se, or as by-products of breeding programs. Loss of material has been excessive, even with close monitoring by interested, dedicated philanthropic organizations. Virtually no systematic studies of the vast collections have been made since the collections were assembled and described, and even fewer results of such studies have been published. The conversion of germplasm banks from last-resort sources of material for plant breeding projects into functional and continuing centers for systematic studies of these vast collections appears to be long overdue.

The other alternatives appear to be loss of the material, which we can hardly afford, or the preservation intact of current sites harboring the various centers of variability of our current and future major crops (Iltis, 1972; Committee on Germplasm Resources, 1978). We do not deny the appeal of the latter alternative, which clearly merits more attention than it has received, but it would need to be accompanied in any case by centers for the study of such materials and germplasm banks for standby storage. An eminent colleague, W. C. Gregory, has vividly delineated the problem in a personal communication: "In his efforts to solve the problem [of germplasm erosion] and to stem the loss, man may be holding a straw against the sea. It appears that the richness and variety of the genetic resources of man's cultivated crops reflect the multiplicity and variety of man's own civilizations both in space and in time. His crops appear to be as wild as he is wild, diverse as he is diverse, opportunistic as he is opportunistic, stable as he is stable, and as uniform as he is uniform. In this view, the forces with which the plant breeder must deal in stemming the loss of genetic resources are the forces of Americanization, Westernization, Europeanization, industrialization, and modernization of the Andean Indian, the denizen of the Amazon, the unique amd multiple peoples of the Caucasus, the Himalayan valleys, and western China, the human interfaces of Nepal, Assam, Abyssinia, and Ethiopia. Thus the voice, 'crying among the weeds,' is being and will be swept away by the winds of the human assault on poverty, starvation, and ignorance--the very things the plant breeder has dedicated his life to alleviate....

"The collection from gene pools and maintenance of germplasm banks in the centers of western-type plant breeding establishments, in tropical gardens, and other contrived facilities may be at best a temporary or stop gap measure, at worst a waste of time and money. The problem will require measures commensurate with the social upheaval working against the survival of genetic resources. My own personal view is that present trends indicate that modern world production technology will expedite the annihilation of such genetic resources as we now have...."

Regardless of the outcome of the discussions concerning the internationally oriented germplasm maintenance centers, the United States needs a frost-free maintenance facility in the tropics or subtropics (Ad hoc Subcommittee of the Agricultural Research Policy Advisory Committee, 1973). The existing facilities at Miami, Florida and Mayaguez, Puerto Rico are inadequate for the above purposes. Many of the U.S. crops are subtropical. Many of their relatives are also perennial and flower only under a short day regime. A living collection of such plants suitable for the dual objectives previously mentioned can be established only in a subtropical, frost-free environment.

The prodigious agricultural production of the United States is based on introduced plants. Some of these plants were from Latin America in Precolumbian times, but the fact remains that none of the important food crops originated in this country. Throughout the world, the major crops are being introduced and promoted on an intensifying scale. It should not be necessary or otherwise desirable to depend so completely on so few crops. There are scores of minor crops which have never been subjected to modern agricultural research. Fifty years ago, it would have been difficult to imagine the present-day importance of sorghum or soybeans. Many of the minor crops would probably respond, just as have soybeans and sorghum, to intensive research efforts. They, too, should be collected and preserved.

Neolithic man began the Agricultural Revolution by domestication, selection, and transportation of plants. He was able to do this because of variability. We, in the midst of sporadic surpluses or of impending shortages and cyclical famines, must maintain what we can of the remaining variation. That variability, maintained in natural gene pools, is valuable stuff. Our legacy to future generations should be adequate germplasm resources from those gene pools, properly maintained, studied, distributed, and replenished.

SUMMARY

A biosystematic rationale is given for the preservation of plant germplasm. The disappearance of genetic resources from natural or agricultural ecosystems, from previous collections of plant material, and from germplasm maintenance operations is noted. A brief history of the collections of indigenous strains of maize and the races of maize in the Americas illustrates the aims, successes, problems, pitfalls, and failures of germplasm maintenance programs. We concluded that the maintenance of existing plant germplasm is imperative; that accessions be maintained individually rather than composited; that germplasm maintenance centers not be considered strictly as service units but be involved in, or allied with, biosystematic research; that adequate tropical or subtropical germplasm maintenance and research facilities are needed for major and minor crops of the United States. Because the backbone of U.S. agriculture is introduced germplasm, it is essential that we collect, maintain, and replenish those genetic resources for our future.

NOTE ADDED IN PROOF

Since this paper was written, two events particularly relevant to the second paragraph of the section MAIZE AND ITS RELATIVES have been published. Tetraploid perennial teosinte has been rediscovered (Guzmán Mejia, R., [1978). Redescubrimiento de *Zea perennis* [Gramineae]. Phytologia 38, 177). A diploid perennial teosinte, morphologically primitive but infertile with maize, has also been found (Iltis, H. H., Doebley, J. F., Guzmán M., R., and Pazy, B. [1979]. *Zea diploperennis* [Gramineae]. A new teosinte from Mexico. Science 203, 186-188).

ACKNOWLEDGMENTS

Some of the details and arrangements pertaining to the maize collections have been obtained at various times by personal communication with: Hermilio Angeles A., Mexico; A. Blumenschein, Brazil; Louis N. Bass, United States; W. L. Brown, United States, Clímaco Cassalett D., Colombia; Mario Gutiérrez G., Mexico and Brazil; E. Hernandez X., Mexico; Howard L. Hyland, United States; Edwin James, United States; Ernesto Paterniani, Brazil; Ricardo Sevilla, Peru; Howard Sprague, United States; E. J. Wellhausen, Mexico. The cooperation and advice of these specialists is gratefully acknowledged. We are also grateful for various committee reports, minutes, and other documents made available to us by the Rockefeller Foundation, U.S. Department of Agriculture, and the Agency for International Development.

Paper no. 5641 of the Journal Series of the N. C. Agricultural Experiment Station.

REFERENCES

Ad hoc Subcommittee of the Agricultural Research Policy Advisory Committee (1973). "Recommended Actions and Policies for Minimizing the Genetic Vulnerability of our Major Crops." A Special Report, Agricultural Research Policy Advisory Committee, The USDA and the Nat. Assoc. State Univs. and Land Grant Colleges Cooperating, Washington, D. C.

Agricultural Research Service (1977). "The National Plant Germplasm System." Program Aid Number 1188, USDA, Washington, D. C.

Allard, R. W. (1965). Genetic systems associated with colonizing ability in predominantly self-pollinating species. *In* "The Genetics of Colonizing Species" (H. G. Baker and G. L. Stebbins, eds.), pp. 49-75. Academic Press, New York.

Allard, R. W., Jain, S. K., and Workman, P. L. (1968). The genetics of inbreeding species. *Advan. Genet. 14,* 55-131.

Baker, H. G. (1971). Commentary. *In* "Man Across the Sea" L. C. Riley, J. C. Kelley, C. W. Pennington, and R. L. Rand, eds.), pp. 428-444. Univ. of Texas Press, Austin.

Beasley, J. O. (1942). Meiotic chromosome behavior in species, species hybrids, haploids, and induced polyploids of *Gossypium. Genetics 27,* 25-54.

Bird, R. McK., and Goodman, M. M. (1977). The races of maize V: grouping maize races on the basis of ear morphology. *Econ. Bot. 34,* 471-481.

Brieger, F. G., Gurgel, J. T. A., Paterniani, E., Blumenschein, A., and Alleoni, M. R. (1958). "Races of Maize in Brazil and other Eastern South American Countries." Nat. Acad. Sci.-Nat. Res. Counc. Publ. 593, Washington, D. C.

Brown, W. L. (1960). "Races of Maize in the West Indies." Nat. Acad. Sci.-Nat. Res. Counc. Publ. 792, Washington, D. C.

Brown, W. L. (1975). Maize germplasm banks in the Western Hemisphere. *In* "Crop Genetic Resources for Today and Tomorrow." (O. H. Frankel, and J. G. Hawkes, eds.), pp. 467-472. Cambridge Univ. Press, Cambridge.

Brown, W. L., and Goodman, M. M. (1977). Races of corn. *In* "Corn and Corn Improvement" (G. F. Sprague, ed.), pp. 49-88. Amer. Soc. Agron., Madison.

Burgess, S. (1971). "The National Program for Conservation of Crop Germ Plasm (A Progress Report on Federal/State Cooperation)." Univ. Printing Dept., Univ. of Georgia, Athens.

Chang, T. T. (1976). The origin, evolution, cultivation and diversification of Asian and African rices. *Euphytica 25*, 425-441.

Chang, T. T., Sharma, S. D., Adair, C. R., and Perez, A. T. (1972). "Manual for Field Collectors of Rice." Internat. Rice Res. Inst., Los Baños.

Chang, T. T., Villareal, R. L., Loresto, G., and Perez, A. T. (1975). IRRI's role as a genetic resource center. *In* "Crop Genetic Resources for Today and Tomorrow" (O. H. Frankel and J. G. Hawkes, eds.), pp. 457-465. Cambridge Univ. Press, Cambridge.

Committee on Genetic Vulnerability of Major Crops. (1972). "Genetic Vulnerability of Major Crops." Nat. Acad. Sci., Washington, D. C.

Committee on Germplasm Resources. (1978). "Conservation of Germplasm Resources: An Imperative." Nat. Acad. Sci., Washington, D. C.

Committee on Preservation of Indigenous Strains of Maize. (1954). "Collections of Original Strains of Maize," Vol. 1. Nat. Acad. Sci.-Nat. Res. Council, Washington, D. C.

Committee on Preservation of Indigenous Strains of Maize. (1955). "Collections of Original Strains of Maize," Vol. 2. Nat. Acad. Sci.-Nat. Res. Council, Washington, D. C.

Cutler, H. C., and Anderson, E. (1941). A preliminary survey of the genus *Tripsacum*. *Ann. Missouri Bot. Garden 28*, 249-269.

de Wet, J. M. J., Gray, J. R., and Harlan, J. R. (1976). Systematics of *Tripsacum* (Gramineae). *Phytologia 33*, 203-227.

de Wet, J. M. J., and Harlan, J. R. (1970). Apomixis, polyploidy, and speciation in *Dichanthium*. *Evolution 24*, 270-277.

Dobzhansky, T. (1951). Mendelian populations and their evolution. *In* "Genetics in the 20th Century" (L. C. Dunn, ed.), pp. 573-589. MacMillan, New York.

Duvick, D. N. (1975). Using host resistance to manage pathogen populations: A corn breeder's commentary. *Iowa State J. Res. 49*, 505-512.

Duvick, D. N. (1977). Major United States crops in 1976. *Ann. N. Y. Acad. Sci. 287*, 86-96.

Food and Agriculture Organization of the United Nations. (1977). *1976 FAO Productions Yearbook, Vol. 30,* Food and Agriculture Organization of the United Nations (FAO), Rome.

Frankel, O. H., and Bennett, E., eds. (1970). "Genetic Resources in Plants--their Exploration and Conservation." Blackwell, Oxford.

Frankel, O. H., and Hawkes, J. G., eds. (1975). "Crop Genetic Resources for Today and Tomorrow." Cambridge Univ. Press, Cambridge.

Galinat, W. C. (1971). The origin of maize. *Ann. Rev. Genetics 5,* 447-478.
Galinat, W. C. (1972). Preserve Guatemalan teosinte, a relict link in corn's evolution. *Science 180,* 323.
Galinat, W. C. (1974). The domestication and genetic erosion of maize. *Econ. Bot. 28,* 31-37.
Galinat, W. C. (1977). The origin of corn. *In* "Corn and Corn Improvement" (G. F. Sprague, ed.), pp. 1-47. Amer. Soc. Agron., Madison.
Gerstel, D. U. (1945). Inheritance in *Nicotiana tabacum* XX. The addition of *Nicotiana glutinosa* chromosomes to tobacco. *J. Hered. 36,* 197-206.
Goodman, M. M., and Bird, R. McK. (1977). The races of maize IV: Tentative grouping of 219 Latin American races. *Econ. Bot. 31,* 204-221.
Grant, U. J., Hatheway, W. H., Timothy, D. H., Cassalett D., C., and Roberts, L. M. (1963). "Races of Maize in Venezuela." Nat. Acad. Sci.-Nat. Res. Counc. Publ. 1136, Washington, D. C.
Grobman, A., Salhuana, W., and Sevilla, R., in collaboration with Mangelsdorf, P. C. (1961). "Races of Maize in Peru." Nat. Acad. Sci.-Nat. Res. Counc. Publ. 915, Washington, D. C.
Hammons, R. O. (1976). Peanuts: genetic vulnerability and breeding strategy. *Crop Sci. 16,* 527-530.
Harlan, J. R. (1972). Genetics of disaster. *J. Environ. Quality 1,* 212-215.
Harlan, J. R. (1975a). "Crops and Man." Amer. Soc. Agron., Madison.
Harlan, J. R. (1975b). Our vanishing genetic resources. *Science 188,* 618-621.
Harlan, J. R., and de Wet, J. M. J. (1971). Toward a rational classification of cultivated plants. *Taxon 20,* 509-517.
Harlan, J. R., Brooks, M. H., Borgaonkar, D. S., and de Wet, J. M. J. (1964). Nature and inheritance of apomixis in *Bothriochloa* and *Dichanthium*. *Botan. Gazette 125,* 41-46.
Harlan, J. R., de Wet, J. M. J., and Price, E. G. (1973). Comparative evolution of cereals. *Evolution 27,* 311-325.
Hatheway, W. H. (1957). "Races of Maize in Cuba." Nat. Acad. Sci.-Nat. Res. Counc. Publ. 453, Washington, D. C.
Hernandez X., E., and Alanis F., G. (1970). Estudio morfologico de cinco nuevas razas de maíz de la Sierra Madre Occidental de Mexico: Implicaciones filogenéticas y fitogeográficas. *Agrociencia 5,* 3-30.
Hernandez X., E., and Randolph, L. F. (1950). "Descripción de los Tripsacum diploides de Mexico: Tripsacum maizar y Tripsacum zopilotense spp. nov." Fol. Tec. No. 4. Ofic. Estud. Esp. Sec. Agr. y Ganad. Mexico, D. F.

Hussaini, S. H., Goodman, M. M., and Timothy, D. H. (1977). Multivariate analysis and the geographical distribution of the world collection of finger millet. *Crop Sci. 17,* 257-263.

Iltis, H. H. (1972). The extinction of species and the destruction of ecosystems. *Amer. Biol. Teacher 34,* 201-205.

Information Sciences/Genetic Resources Program. (1977). "Maize Directory--1977." IS/GR, Univ. Colorado, Boulder.

International Board for Plant Genetic Resources (1975). "The Conservation of Crop Genetic Resources." Whitefriars Press, London.

International Board for Plant Genetic Resources (1976). "Priorities Among Crops and Regions." Food and Agriculture Organization of the United Nations (FAO), Rome.

Kashiwakura, Y., and Paterniani, E. (1972). O armazenamento de sementes de milho em dessecador com silica gel e em câmera seca, e o seu efeito na preserva ção da viabilidade de sementes de milho. Universidade de São Paulo Escola Superior de Agricultura "Luiz de Queiroz," Instituto de Genetica. *Relatorio Cientifico 6,* 38-48.

Levings, C. S. III, and Pring, D. R. (1976). Restriction endonuclease analysis of mitochondrial DNA from normal and Texas cytoplasmic male sterile maize. *Science 193,* 158-160.

Levings, C. S. III, and Pring, D. R. (1977). Diversity of mitochondrial genomes among normal cytoplasms of maize. *J. Hered. 68,* 350-354.

Mangelsdorf, P. C. (1966). Genetic potentials for increasing yields of food crops and animals. *Proc. Nat. Acad. Sci. (U.S.) 56,* 370-375.

Mangelsdorf, P. C. (1974). "Corn: Its Origin, Evolution and Improvement." Harvard Univ. Press, Cambridge.

Mangelsdorf, P. C., MacNeish, R. S., and Galinat, W. C. (1964). Domestication of cron. *Science 143,* 538-545.

Mangelsdorf, P. C., MacNeish, R. S., and Galinat, W. C. (1967). Prehistoric wild and cultivated maize. *In* "The Prehistory of the Tehuacan Valley, Volume One. Environment and Subsistence." (D. S. Byers, ed.), pp. 178-200. Univ. of Texas Press, Austin.

Marshall, D. R., and Allard, R. W. (1970). Maintenance of isozyme polymorphisms in natural populations of *Avena barbata*. *Genetics 66,* 393-399.

Oka, H., and Morishima, H. (1971). The dynamics of plant domestication: Cultivation experiments with *Oryza perennis* and its hybrid with *O. sativa*. *Evolution 25,* 356-364.

Paterniani, E., and Goodman, M. M. (1978). "Races of Maize in Brazil and Adjacent Areas." International Maize and Wheat Improvement Center (CIMMYT), Mexico City.

Pring, D. R., and Levings, C. S. III. (1978). Heterogeneity of maize cytoplasmic genomes among male-sterile cytoplasms. *Genetics 89,* 121-136.

Pring, D. R., Levings, C. S. III., Hu, W. W. L., and Timothy, D. H. (1977). Unique DNA associated with mitochondria in the "S" type cytoplasm of male sterile maize. *Proc. Nat. Acad. Sci. (U.S.) 74,* 2904-2908.

Ramírez E., R., Timothy, D. H., Díaz B., E., and Grant, U. J., in collaboration with Nicholson C., G. E., Anderson, E., and Brown, W. L. (1960). "Races of Maize in Bolivia." Nat. Acad. Sci.-Nat. Res. Counc. Publ. 747, Washington, D. C.

Randolph, L. F. (1970). Variation among *Tripsacum* populations of Mexico and Guatemala. *Brittonia 22,* 305-307.

Roberts, L. M., Grant, U. J., Ramírez E., R., Hatheway, W. H., and Smith, D. L., in collaboration with Mangelsdorf, P. C. (1957). "Races of Maize in Colombia." Nat. Acad. Sci.-Nat. Res. Counc. Publ. 510, Washington, D. C.

Sears, E. R. (1956). The transfer of leaf-rust resistance from *Aegilops umbellulata* to wheat. *In* "Genetics in Plant Breeding." Brookhaven Symposia in Biology *9,* 1-22.

Smith, H. H. (1971). Broadening the base of genetic variability in plants. *J. Hered. 62,* 265-276.

Stebbins, G. L. (1950). "Variation and Evolution in Plants." Columbia Univ. Press, New York.

Timothy, D. H. (1972). Plant germplasm resources and utilization. *In* "The Careless Technology: Ecology and International Development." (M. T. Farvar, and J. P. Milton, eds.), pp. 631-656. Natural History Press, Garden City, N. J.

Timothy, D. H., Pẽna V., B., and Ramírez E., R., in collaboration with Brown, W. L., and Anderson, E. (1961). "Races of Maize in Chile." Nat. Acad. Sci.-Nat. Res. Counc. Publ. 847, Washington, D. C.

Timothy, D. H., Hatheway, W. H., Grant, U. J., Torregroza C., M., Sarria V., D., and Varela A., D. (1963). "Races of Maize in Ecuador." Nat. Acad. Sci.-Nat. Res. Counc. Publ. 975, Washington, D. C.

Wade, N. (1972). A message from corn blight: The dangers of uniformity. *Science 177,* 678-679.

Webster, O. J. (1976). Sorghum vulnerability and germplasm resources. *Crop Sci. 16,* 553-556.

Wellhausen, E. J., Roberts, L. M., and E. Hernandez X., in collaboration with P. C. Mangelsdorf. (1952). "Races of Maize in Mexico." Bussey Institution, Harvard Univ., Cambridge.

Wellhausen, E. J., Fuentes O., A., and Hernández Corzo, A., in collaboration with Mangelsdorf, P. C. (1957). "Races of Maize in Central America." Nat. Acad. Sci.-Nat. Res. Counc. Publ. 511, Washington, D. C.

Wilkes, H. G. (1967). "Teosinte: The Closest Relative of Maize." Bussey Institution, Harvard Univ., Cambridge.

Wilkes, H. G. (1972a). Maize and its wild relatives. *Science 177*, 1071-1077.

Wilkes, H. G. (1972b). Genetic erosion in teosinte. *Plant Genet. Resources News. 28*, 3-10.

Wilkes, H. G. (1977). Hybridization of maize and teosinte, in Mexico and Guatemala and the improvement of maize. *Econ. Bot. 31*, 254-293.

Wilkes, H. G., and Wilkes, S. (1972). The green revolution. *Environment 14*, 32-39.

Seed Preservation

NUCLEOTIDE METABOLISM AND THE GERMINATION OF SEED EMBRYONIC AXES[1]

Shirley Rodaway
Bor-Fuei Huang
Abraham Marcus

The Institute for Cancer Research
The Fox Chase Cancer Center
Philadelphia, Pennsylvania

The quiescent state of the mature plant seed is characterized by arrested growth, suspended development, and a subsistence level of metabolism (Bewley and Black, 1978; Khan, 1977). With the proper environmental stimulus, biochemical processes that were markedly decreased during seed maturation readjust to a state more characteristic of a growing system. These metabolic processes are not all stimulated simultaneously but are activated either in series or in parallel, depending on the mechanisms responsible for their control.

The studies to be described consider that hydration of the embryonic axis of the dry seed stimulates axis metabolism by a series of progressive changes in specific biochemical processes. The axis can be excised from the rest of the dry seed, allowing a more precise study of the biochemistry that accompanies its germination in the absence of events occurring in other parts of the seed such as the cotyledons or aleurone tissue.

As part of a long-term study on the biochemistry of germination of the embryonic axis of the wheat grain, we have become interested in the role of increasing levels of ATP and GTP with regard to increases in the rate of protein synthesis seen during the early hours of germination. We have further

[1]*This work was supported by grant PCM75-18878 from the National Science Foundation; by U.S.P.H.S. grants CA-06927 and RR-05539 from the National Institutes of Health; and by an appropriation from the Commonwealth of Pennsylvania.*

ISBN 0-12-602050-7

extended our observations and conclusions to the embryonic axis of soybeans. We have considered the possibility that synthesis of certain other nucleotides may be required for growth of the embryonic axis. As a result, the second part of this article considers the relationship between growth of the soybean embryonic axis and the metabolism *in vivo* of a major group of nucleotides, the nucleotide sugars. The nucleotide sugars are required for the synthesis of the plant cell wall (Karr, 1976; Maclachlan, 1977) and the oligosaccharide moieties of glycoproteins (Brown and Kimmins, 1977), as well as for other anabolic reactions of monosaccharides. Control of their metabolism could therefore provide regulatory mechanisms for the initiation of growth by the axis.

ADENYLATES, GUANYLATES, AND RATES OF PROTEIN SYNTHESIS

Isolated embryonic axes from dry wheat and soybean seeds increase their fresh weight upon exposure to water (Marcus *et al.*, 1975). The wheat axes become fully hydrated to a little

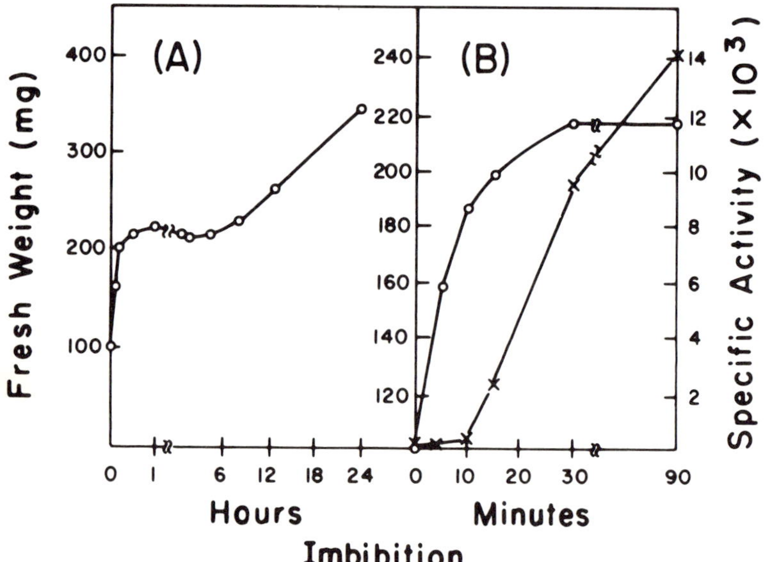

FIGURE 1. Fresh weight changes in wheat embryonic axes during early germination. Isolated wheat axes (100 mg) were imbibed, then blotted and weighed. (B) Fresh weight determinations were made and the ribosomal fraction was assayed for ability to direct protein synthesis in vitro (Marcus et al., 1975).

over twice their initial weight after about 20 min (Fig. 1A). This initial period of water gain is largely a physical process, since heat-killed embryos also take up water in this way. Subsequent to the initial water uptake, there is a period of about 5 hr during which time the axis is quiescent with respect to growth. We refer to this period as the quiescent phase. Thereafter the fresh weight of the axis begins to increase linearly. Figure 1B shows an expansion of the first part of the fresh weight curve in Fig. 1A. During this early period of incubation, one can measure the ability of the wheat ribosomal fraction to direct the incorporation of amino acids into proteins *in vitro*. Ribosomal fractions from the dry embryos are essentially inactive, whereas similar fractions, taken from axes at 10-15 min after the onset of imbibition, are considerably more active.

This increased capacity for protein synthesis is gained at the same time that poly(A)+RNA, already present in the dry embryos, becomes associated with the free ribosomes to form polysomes (Marcus *et al.*, 1975; Brooker *et al.*, 1977). Association of the poly(A)+RNA with the polyribosomes is maximal by the beginning of the quiescent period (Spiegel and Marcus, 1975) and experiments using cDNA prepared using poly(A)+RNA as template have shown that the predominant kinds of poly(A)+RNA in the embryonic axes of wheat differed little in dry, 45 min, and 5.5 hr imbibed axes (Brooker *et al.*, 1978).

The increase in the rate of protein synthesis, although most dramatic in the initial 40 min of imbibition, occurs also during the quiescent period. Such rate changes can be measured

TABLE I. Rates of Protein Synthesis in Wheat Embryonic Axes during Early Germination[a]

Germination	*^{3}H-Leucine incorporation (cpm)*	
	TCA-soluble	*TCA-insoluble*
40 min	*106,215*	*2333 (1.0)*
3 hr	*120,235*	*5827 (2.5)*
5.5 hr	*152,000*	*7920 (3.4)*
8 hr	*204,000*	*9330 (4.0)*

[a]*Wheat embryonic axes (125mg) were imbibed as indicated, then incubated for 10 min in 8.3 mM ^{3}H-leucine. The washed embryos were homogenized in cold 5% TCA, and the TCA-soluble and insoluble radioactivities were determined (Brooker* et al., *1977).*

quantitatively *in vivo* by determining the rate of incorporation of ^{3}H-leucine into protein in intact embryos (Table I). Between 40 min and 5.5 hr of germination, the rate of protein synthesis more than triples. Thereafter, when growth begins (in the interval between 5.5 and 8 hr) protein synthesis increases only slightly. The date of Table II lend support to our not correcting the data of Table I for differential uptake of label by the axes. Low exogenous concentrations of leucine exaggerate the difference in protein synthesis rates. As the leucine concentration is increased, the apparent rate of ^{3}H-

TABLE II. Effect of Amino Acid Concentration on Incorporation of ^{3}H-Leucine into Protein[a]

Leucine concentration	Germination	^{3}H-Leucine incorporation, cpm	
		TCA-soluble	TCA-insoluble
(1) 0.3×10^{-6} M	40 min	121,500	4480 (1.0)
	3 hr	115,830	36,000 (8.0)
	5.5 hr	144,630	69,840 (15.6)
(2) 8.3×10^{-4} M	40 min	73,710	3437 (1.0)
	3 hr	87,480	20,190 (5.9)
	5.5 hr	114,450	33,959 (9.9)
(3) 8.3×10^{-3} M	40 min	77,190	1910 (1.0)
	3 hr	90,560	5246 (2.7)
	5.5 hr	102,140	8784 (4.6)

[a]10 min assays of ^{3}H-leucine incorporation were performed and analyzed as in Table I, except that the specific activity of the radioactive leucine was varied as follows: (1) ^{3}H-leucine at 12,741 cpm/pmol, (2) 5,095 cpm/pmol, and (3) 510 cpm/pmol. The figures in parentheses are rates of incorporation relative to a value of 1.0 for the 40 min sample (Cheung, Huang, and Marcus, unpublished data).

leucine incorporation into protein becomes less of a reflection of the specific activity of ^{3}H-leucine in the tissue; rather, it reflects the specific activity in the medium. As a result, the rate of incorporation of radioactivity into protein is independent of the amount of label taken up by the axes when the leucine concentration is 8.3 mM.

At this point we asked, what biochemical events could be responsible for these changes in the rates of protein synthesis? Early work on *in vitro* protein synthesis had already shown that ATP was a necessary component of the reaction mixture (Marcus and Feeley, 1966). We therefore postulated that the steady state levels of ATP in the tissue might regulate the rate of protein synthesis in the wheat embryos. That is, the level of ATP in the tissue might affect the rate of protein synthesis if ATP was a substrate for translation of mRNA and if the level of ATP in the tissue was reasonably below saturation for these reactions.

As shown in Table III, dry embryos contain little ATP, most of the adenylates being in the form of AMP (Obendorf and Marcus, 1974). During hydration, the AMP is rapidly converted, presumably through ADP, to ATP. Within the imbibition period, the level of ATP rises well over 100-fold. Because ATP is required for translation *in vitro* (Marcus and Feeley, 1966), it seemed possible that the initial increase in the rate of protein synthesis that occurs during the imbibition period indeed might be related to the large increase in the steady state level of ATP. Between 40 min and 5.5 hr, however, the ATP level increases

TABLE III. Levels of Adenine Nucleotides in Wheat Embryonic Axes during Early Germination[a]

	ATP	ADP	AMP	Total	Adenylate charge
	(nmoles/125 mg embryos)				
Dry	1	47	159	207	0.12
20 min	133	26	13	172	0.85
40 min	161	23	5	189	0.91
3 hr	250	23	8	281	0.93
5.5 hr	259	24	3	286	0.95

[a]*Wheat axes (125 mg) were imbibed as indicated, homogenized in 5% TCA, and assayed for the presence of the indicated adenine nucleotides (Brooker et al., 1977).*

only another 60%; yet recall that in that same period the rate of protein synthesis more than triples. In the quiescent period, therefore, either protein synthesis is extremely sensitive to small changes in the levels of ATP or some other mechanism is operating that further regulates the rate of protein synthesis.

The wheat embryonic axes used in these experiments are isolated mechanically, and the axes are often broken with an occasional piece of the scutellum being retained. Soybean embryonic axes could be isolated manually and separated completely from their cotyledons. When exposed to water, the increase in fresh weight of these axes, as with the wheat embryos, was found to be triphasic (Fig. 2). With the soybean system the "quiescent" period extends from 90 min after H_2O addition until 10 hr. During this period the rate of protein synthesis, as measured by incorporation of ^3H-leucine into protein *in vivo*, increases 13-fold (Table IV), whereas the ATP level increases only 40% (Table V). Qualitatively, the results with the soybean and wheat embryonic axes are the same in that both exhibit a striking increase in the level of ATP during imbibition, and a further but less dramatic increase at the later time when protein synthesis rates are increasing. In soybean axes, however, the increase in the rate of protein synthesis that occurs during the quiescent period far exceeds the increase that would be expected from a 40% increase in the ATP level. These data, therefore, lead to the conclusion that while the rise in protein synthesis during the imbibition phase may require an increase in the steady state level of ATP, other mechanisms must function during the quiescent period to further increase the rate of protein synthesis.

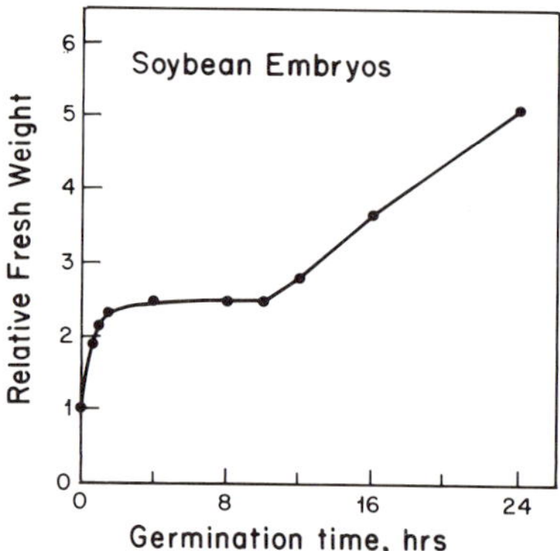

FIGURE 2. Fresh weight changes in isolated soybean embryonic axes during early germination. Soybean embryos were imbibed for the times shown, blotted, and weighed. The initial fresh weights were about 130 mg per 30 axes.

TABLE IV. Rates of Protein Synthesis by Soybean Embryonic Axes during Early Germination[a]

Germination	^{3}H-Leucine incorporation (cpm) TCA-soluble	TCA-insoluble
40 min	47,080	187 (1.0)
90 min	34,510	301 (1.6)
4 hr	49,355	2,063 (11)
10 hr	60,950	3,919 (21)
16 hr	57,640	4,758 (26)

[a]*Thirty soybean axes were imbibed as indicated, and the rates of protein synthesis were measured in vivo for a 10 min period as in Table I. ^{3}H-Leucine (8.3 mM, 2 μCi) was used in a volume of 1.6 ml.*

TABLE V. Levels of Adenine Nucleotides in Soybean Embryonic Axes during Early Germination

	ATP	ADP	AMP	Total	Adenylate charge
	(nmoles per embryo)				
Dry	0.1	1.1	1.8	3.0	0.21
40 min	3.3	1.0	0.3	4.7	0.82
90 min	4.7	1.1	0.3	6.0	0.86
4 hr	6.3	0.9	0.3	7.6	0.89
10 hr	6.5	0.8			
16 hr	7.9	1.1			

[a]*Soybean embryonic axes were imbibed as indicated, homogenized in 5% TCA, and analyzed for adenine nucleotides according to Cheung and Marcus (1975).*

The increased rate of protein synthesis also occurs independently of changes in the level of GTP. As with ATP, GTP is another nucleotide required for translation of mRNA *in vitro*. In wheat embryos, the level of GTP rises predominantly during the initial imbibition period, with at most only a 20% increase occurring between 40 min and 5.5 hr (Table VI). Likewise, in soybean there is only a 30% increase in the level of GTP be-

TABLE VI. Levels of Guanine Nucleotides in Wheat Embryonic Axes during Early Germination[a]

	GTP	GDP	GMP	Total
	(nmoles/125 mg embryos)			
Dry	6	16	12	34
40 min	40	4	4	48
3 hr	38	5	4	47
5.5 hr	53	6	3	62

[a]*Wheat embryonic axes (125 mg) were imbibed in H_2O, homogenized in 5% TCA, and assayed for guanine nucleotides (Brooker et al., 1977). The procedure used (Cheung and Marcus, 1976) yields values for GTP which include the GDP-sugars, for which a correction can be made. The corrected GTP levels are 5.2, 37, 35, and 48 nmol, respectively.*

tween 90 min and 10 hr (Table VII). Thus changes in the amount of GTP in the embryonic axes during the quiescent period do not account for the much larger changes in the rates of protein synthesis as measured *in vivo*.

TABLE VII. Levels of Guanine Nucleotides in Soybean Embryonic Axes during Early Germination[a]

	GTP	GDP+GMP	Total
	(nmoles per embryo)		
Dry	0.11	0.68	0.8
40 min	0.51	0.55	1.1
90 min	0.95	0.55	1.5
4 hr	0.90	0.48	1.4
10 hr	1.32	0.45	1.8
16 hr	1.48		

[a]*Extraction of 30 soybean axes was performed after the indicated incubation periods. Guanine nucleotides were determined as described (Cheung and Marcus, 1976), with a correction made for the reaction of GDP-sugars in the GTP assay.*

NUCLEOTIDE SUGAR METABOLISM

Elongation growth of embryonic axes requires changes in the structure of the cell wall. One way in which the wall structure could be changed would be by addition of oligosaccharides to the wall matrix. *In vitro* studies have shown that oligosaccharides similar to those in the cell wall are synthesized from monosaccharides "activated" through covalent linkage to nucleoside diphosphates (Karr, 1976). In addition, oligosaccharides moieties of glycoproteins are also formed using nucleoside diphosphate sugars (NDP-sugars) as monosaccharide donors (Lennarz, 1975; Brown and Kimmins, 1977). Thus a possible control point for regulating growth is the ability of the axis to synthesize a specific class of NDP-sugars.

Dry soybean axes contain about 1.1 nmole of UDP-sugars per embryo (Fig. 3). Upon hydration more UDP-sugars are synthesized, and the level of these compounds increases steadily throughout all three phases of germination. This steady rise in the level of UDP-sugars, rather than an abrupt increase just prior to the initiation of growth, suggests that most enzymes and substrates being utilized for UDP-sugar synthesis during the early growth period (10-16 hrs) are available during

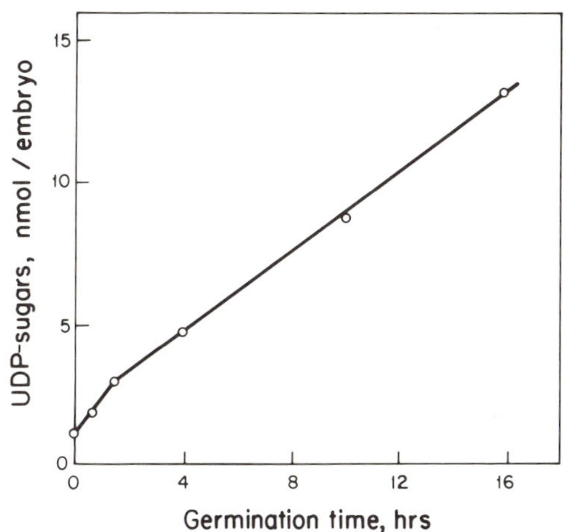

FIGURE 3. Levels of total UDP-sugars in soybean embryonic axes during early germination. Cold 5% TCA extracts were treated with ether and neutralized. The level of UDP-sugars was determined by a coupled enzyme system that used sequential treatments with alkaline phosphatase, nucleotide pyrophosphatase, and nucleoside monophosphate kinase to generate UDP from UDP-sugars. The amount of UDP was then determined indirectly by measuring the amount of ADP (Cheung and Marcus, 1975) formed from ATP in the reaction with NMP kinase (manuscript in preparation).

the 10 hr period before growth is initiated. The kinetics of GDP-sugar accumulation suggest that a similar situation occurs during the synthesis of these compounds (Fig. 4). During imbibition there is a rapid increase in the level of total GDP-sugars, with a transient decrease during the early part of the quiescent phase and an increase thereafter. The level of GDP-sugars at 10 hr is not much greater than at 90 min, so that the level of total GDP-sugars cannot be regulating the onset of growth.

Synthesis of nucleoside diphosphate sugars requires a number of enzymes for synthesizing specific sugar phosphates and their respective sugar nucleotides, as well as for interconverting sugar moieties in the form of NDP-sugars. Considering that the synthesis of a specific sugar nucleotide might be limiting to growth, we labeled the nucleoside moieties of the UDP- and GDP-sugars in a 60 min exposure to either ^{3}H-uridine or ^{3}H-guanosine. We then resolved the nucleotide sugars in a number of fractionation procedures.

Tables VIII and IX contain the analyses of the labeling patterns of ^{3}H-UDP sugars. The time points of 4, 9, and 15 hrs were chosen to represent early quiescence, the end of quiescence, and the early growth phase. By fractionating on DEAE-cellulose, three UDP-sugar fractions are obtained (Table VIII). The distribution of label in the three fractions is such that the labeling of UDP-uronic acids increases about 20% with the

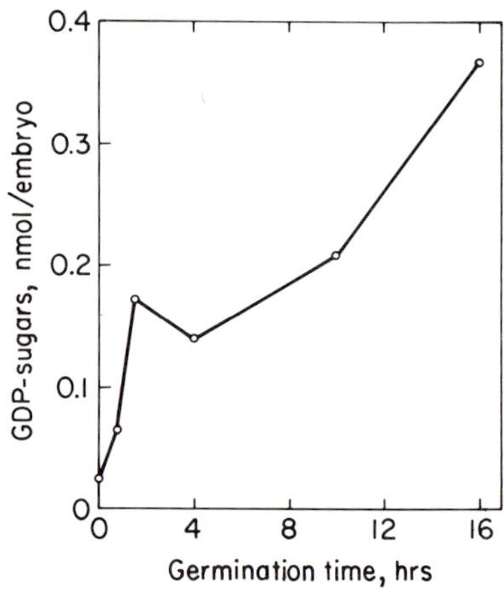

FIGURE 4. Levels of total GDP-sugars in soybean embryonic axes during early germination. Nucleotide sugars were extracted as in Fig. 3, then the level of GDP-sugars was determined by a coupled enzyme system. After alkaline phosphatase treatment, the extract was treated with nucleotide pyrophosphatase and assayed for GMP, essentially using the procedures referred to earlier (Cheung and Marcus, 1976).

TABLE VIII. Distribution of Radioactivity in UDP-Sugar Fractions after a 60 Minute Labeling Period[a]

Peak	Total radioactivity (%)		
	4 hr	9 hr	15 hr
A	2.5	2.6	4.3
B	93.2	93.0	90.5
C	4.3	4.3	5.2

[a]*Soybean embryonic axes were incubated as indicated, then were labeled for 60 min with* ^3H*-uridine. Axes were extracted with cold 5% TCA, treated with alkaline phosphatase, then chromatographed on columns of Bio-Gel P2 and Whatman DE 52 which yielded three separate peaks of UDP-sugars. Peak A contained unknown UDP-sugars, peak B the UDP-hexoses and UDP-NAc hexosamines, and peak C the UDP-uronic acids.*

TABLE IX. Distribution of Radioactivity in UDP-Sugar Classes after a 60 Minute Labeling Period[a]

Class	Comigrating marker	Total radioactivity (%)		
		4 hr	9 hr	15 hr
I	(origin)	0.009	-	0.007
II	none	0.29	0.75	4.13
III	UDP-galUA	0.38	0.70	0.59
IV	UDP-glcUA	3.34	2.91	3.48
V	none	0.61	0.60	1.01
VI	UMP	0.47	0.45	2.0
VII	UDP-gal,(-man)	21.3	22.8	15.8
VIII	UDP-glc,(-xyl)	62.3	62.3	59.0
IX	UDP-glcNAc	8.51	6.74	9.75
X	none	2.34	2.41	4.17

[a]*Soybean embryonic axes were incubated for the indicated times plus 60 min in* ^3H*-uridine. Nucleotide sugars were fractionated as indicated on Table VIII, then analyzed by thin layer chromatography on PEI-cellulose (Randerath and Randerath, 1965, 1966). Classes originate from the following DE52 fractions: A(X); B(II, VI-IX); C(I, III-V).*

initiation of growth, while the level of an unidentified fraction (less negatively charged than the other two fractions) increases 80% at that time. The UDP-sugar fractions were further resolved into several distinct classes of UDP-sugars by thin layer chromatography (Table IX). Commercially available sugar nucleotides were used as markers nominally to identify most of these classes, although other UDP-sugars of similar structure may migrate to some of the same positions. The major UDP-sugars labeled are neutral sugar derivatives, and the distribution of label into these compounds varies little. The three groups of unknown compounds show changes in labeling efficiency, but the significance of this is unclear. Overall, the data suggest that there are no abrupt changes in the relative synthesis of UDP-sugars as growth begins, and that even in growing axes there are likewise only minor changes in the labeling patterns of the major uridine-sugar nucleotides. When ^{3}H-guanosine was used to label the GDP-sugars, three classes of GDP-sugars could be separated (Table X). Again there was no major shift in the distribution of radioactivity in the GDP-sugars between the early quiescent stage and growing embryonic axes.

The UDP-neutral sugar fractions could include UDP-hexoses, UDP-pentoses, and UDP-deoxyhexoses of the galactose, mannose, and glucose families (Khym, *et al.*, 1959). Attempts to resolve these types of compounds directly from the appropriate uridine-labeled fractions (Table IX, classes VII and VIII) were unsuccessful. In an alternative approach, we labeled 16 hr imbibed axes with ^{14}C-glucose from 11 hr to 16 hr in addition to the usual ^{3}H-uridine labeling at 15 hr. Assuming that the ^{14}C radioactivity would be randomized among the different UDP-sugar

TABLE X. Distribution of Radioactivity in GDP-Sugars after a 60 Minute Labeling Period[a]

	Total radioactivity (%)		
	4 hr	*9 hr*	*15 hr*
unknown	*6*	*5*	*6*
GDP-mannose	*50*	*59*	*45*
GDP-hexNAc	*42*	*35*	*50*

[a]*Soybean embryonic axes were incubated and extracts were analyzed as in Tables VIII and IX, except that ^{3}H-guanosine was used to label radioactively the GDP-sugars and the P2 step was omitted.*

moieties, we would have, at the least, a qualitative method for ascertaining the synthesis of minor components, that is, by examining the sugar moieties after hydrolysis. Surprisingly, only the UDP-glc and UDP-glcNAc classes (Table IX, classes VIII and IX) were efficiently labeled by ^{14}C-glucose (Fig. 5). The ^{14}C/^{3}H (cpm) ratios for UDP-gal (class VII), UDP-glc (class VIII) and UDP-glcNAc (class IX) were 0.02, 0.16 and 0.18 respectively, while the ratio of label taken up by the axes was about 0.45. No distinct labeling of UDP-galUA or UDP-glcUA by ^{14}C-glucose could be seen. In addition, the distribution of ^{14}C across the ^{3}H-labeled UDP-glc peak (class VIII), was shifted to one side, suggesting the presence of two uridine labeled compounds, only one of which contains the glucose radioactivity. Authentic UDP-^{14}C xylose chromatographs with the more mobile side of the ^{3}H peak, suggesting that UDP-xylose is efficiently

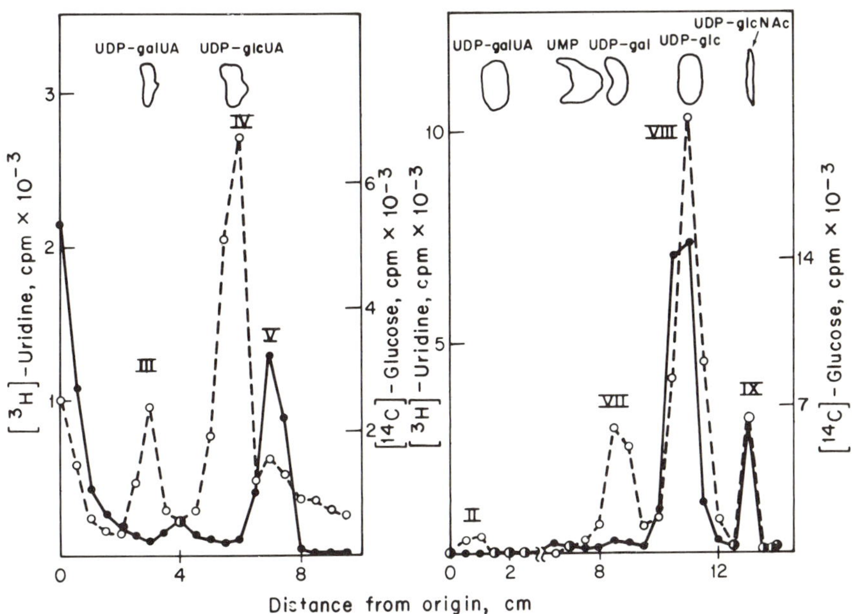

FIGURE 5. Double labeling of 16 hr soybean embryonic axes. Axes were incubated a total of 16 hr, ^{14}C-glucose being added at 11 hr and ^{3}H-uridine at 15 hr. Extraction and separation of the nucleotides is explained in Table IX. Samples were corrected for cross-over during scintillation counting. ^{3}H-uridine, dashed line; ^{14}C-glucose, solid line.

labeled by uridine but not by glucose. Tentatively, therefore, it would appear that some of the UDP-sugars are synthesized from stored sugars or sugar precursors that are not metabolized to glucose or glucose phosphate. Due to the poor labeling of individual UDP-sugars by ^{14}C-glucose, however, the method cannot be used for elucidating the minor UDP-sugars.

The ability of embryonic axes to synthesize the major classes of UDP-sugars does not appear to be the stimulus for growth since these classes are being synthesized as early as the beginning of the quiescent period. There are minor changes in distribution of label within the UDP-sugars, and these may reflect subtle changes in the kinds of glycoproteins and cell wall polymers being synthesized by the cell. The transient decrease in the level of GDP-sugars during the quiescent phase could correlate with the synthesis of macromolecules that may have roles in early growth. The most striking observation is the large accumulation of UDP-sugars, and we suggest that this may in some manner contribute to the initiation of axis growth at 10 hr. It remains, however, to be shown that growth and UDP-sugar synthesis are interconnected events.

SUMMARY

Embryonic axes of soybean and wheat seeds were used to study the sequential activation of biochemical processes during the first few hours of germination. Early germination can be described by three phases: imbibition, a quiescent period, and a growth period. During the imbibition period the levels of ATP and GTP are rapidly increased, predominantly at the expense of stored AMP and GMP. At nearly the same time, ribosome activity is stimulated, suggesting a close relation between these events. During the quiescent period there is a further marked increase in the rate of protein synthesis. In this period, however, there are only minor increases in the levels of ATP and GTP.

Other nucleotides may have important roles in early germination. The nucleotide sugars, as monosaccharide donors during cell wall and oligosaccharide synthesis, potentially could be rate limiting to the growth of embryonic axes. Analysis of these compounds showed the levels of nucleotide sugars to increase during all three germination phases with only minor changes in the distribution of radioactivity (from ^{3}H-uridine or ^{3}H-guanosine) in either the UDP-sugars or the GDP-sugars. These results open the possibility that it may be the cumulative level of the UDP-sugars, rather than the specific synthesis of a particular UDP- or GDP-sugar, that may have a direct relation to the initiation of growth.

REFERENCES

Bewley, J. D., and Black, M. (1978). "Physiology and Biochemistry of Seeds in Relation to Germination," Vol. 1., Springer-Verlag, New York.
Brooker, John C., Cheung, C. P., and Marcus, A. (1977). The physiology and biochemistry of seed dormancy and germination. *In* "Protein Synthesis and Seed Germination" (A Khan, ed.), pp. 347-356. Elsevier/North-Holland Biomedical Press, New York.
Brooker, J. D., Tomaszewski, M., and Marcus, A. (1978). preformed messenger RNAs and early wheat embryo germination. *Plant Physiol. 61,* 145-149.
Brown, R. G., and Kimmins, W. C. (1977). Glycoproteins. *In* "International Review of Biochemistry, Vol. 13" (D. H. Northcote, ed.), pp. 183-208. Univ. Park Press, Baltimore.
Cheung, C. P., and Marcus, A. (1975). Analysis of adenine nucleotides at the picomole level with ^{32}P phosphoenol pyruvate and pyruvate kinase. *Anal. Biochem. 69,* 131-139.
Cheung, C. P., and Marcus, A. (1976). Guanine nucleotide determination in extracts of wheat embryo. *FEBS Letters 70,* 141-144.
Khan, A. A. (1977). "The Physiology and Biochemistry of Seed Dormancy and Germination." North Holland Publishing Company, New York.
Karr, A. L. (1976). Cell wall biogenesis. *In* "Plant Biochemistry, 3rd ed." (J. Bonner and J. E. Varner, eds.), pp. 405-426.
Khym, J. X., Zill, L. P., and Cohn, W. E. (1959). Separation of carbohydrates. *In* "Ion Exchangers in Organic and Biochemistry" (C. Calmon and T. R. E. Kressman, eds.). Interscience Publishers, New York.
Lennarz, W. J. (1975). Lipid linked sugars in glycoprotein synthesis. *Science 188,* 986-991.
Maclachlan, G. (1977). Cellulose metabolism in growing cells. *Trends in Biochem. Sci. 2,* 226-228.
Marcus, A., and Feeley, J. (1966). Ribosome activation and polysome formation *in vitro*: requirement for ATP. *Proc. Nat. Acad. Sci. U.S. 56,* 1770-1777.
Marcus, A., Spiegel, S., and Brooker, J. D. (1975). Preformed mRNA and the programming of early embryo development. *In* "Control Mechanisms in Development" (R. H. Meints and E. Davies, eds.), pp. 1-19. Plenum Publishing Company, New York.
Obendorf, R. L., and Marcus, A. (1974). Rapid increase in adenosine 5'-triphosphate during early wheat embryo germination. *Plant Physiol. 53,* 779-781.

Randerath, K., and Randerath, E. (1965). Ion-exchange thin-layer chromatography. XIV. Separation of nucleotide sugars and nucleoside monophosphates on PEI-cellulose. *Anal. Biochem. 13*, 575-579.

Randerath, K., and Randerath, E. (1966). Ion-exchange thin-layer chromatography. XV. Preparation, properties, and applications of paper-like PEI-cellulose sheets. *J. Chromatogr. 22*, 110-117.

Spiegel, S., and Marcus, A. (1975). Polyribosome formation in early wheat embryo germination independent of either transcription or polyadenylation. *Nature 256*, 228-230.

DORMANCY BREAKING BY HORMONES AND OTHER CHEMICALS--ACTION AT THE MOLECULAR LEVEL[1]

J. Derek Bewley

Department of Biology
University of Calgary
Calgary, Alberta, Canada

Several studies have been initiated to elucidate the mechanism whereby applied hormones and other chemicals release seeds from their dormant condition. Since the molecular events specifically associated with the triggering of germination *per se* are as yet unknown, perhaps it is not surprising that the studies so far have given us very little insight into the mode of action of hormones and related substances at the molecular level. In this article, essentially a review, I hope to show that even less is known about their action than is often assumed. Space does not permit a comprehensive review and my intention is to highlight certain experiments and experimental approaches, their strengths and their weaknesses. A longer and more in-depth account can be found in Black and Bewley, 1979.

I will confine myself to discussing the possible action of hormones and chemicals during germination, which includes those events occurring between initial imbibition and radicle protrusion through the surrounding seed structures. Events occurring thereafter are concerned with growth, and not germination. Despite occasional statements to the contrary in the literature, the breaking of dormancy and stimulation of radicle emergence is not mediated through promotion of hydrolysis of reserves. This is a postgermination event and is associated with, if not vital for, establishment of the seedling (Bewley and Black, 1978).

[1]*Supported by National Research Council of Canada grant A6352 and an appropriation from the University of Calgary grants committee. The author was the holder of a Killam Resident Fellowship during the preparation of this manuscript.*

ISBN 0-12-602050-7

In the absence of any concrete understanding of the molecular events associated with germination, workers have concentrated on the effects of hormones that are accepted to be prerequisite for successful radicle emergence (e.g., nucleic acid and protein synthesis, respiratory metabolism). For it has been suggested that dormancy could result from some impediment to one or more of these processes. It is important to note that dormancy should not be equated with metabolic quiescence, for there is ample evidence that dormant seeds are metabolically active, e.g., dormant lettuce seeds respire (Woodstock and Toole, 1976) and conduct RNA (Frankland *et al.*, 1971) and protein synthesis (Fountain and Bewley, 1973). Thus any promotive action of hormones is likely to entail some modification or enhancement of this "basal" metabolism.

HORMONAL EFFECTS ON NUCLEIC ACID AND PROTEIN SYNTHESIS

On the premise that germination may require the synthesis of specific enzymes/proteins, it has been of interest to determine if applied gibberellin can modulate transcription and/or translation to bring about quantitative and qualitative shifts in protein synthesis. To date, the results obtained are not too revealing.

A considerable body of work has been carried out to determine the effects of RNA and protein synthesis inhibitors on GA-induced germination. But the results obtained have been very variable, the effects of the inhibitors ranging from completely inhibitory to completely ineffective. Conclusions drawn from inhibitor studies alone should be treated with due caution, for the following reasons: (*a*) Proof is often lacking that inhibitors are taken up by the seed. Thus claims, for example, that germination occurring in the presence of RNA synthesis inhibitors is independent of new RNA synthesis must be treated with skepticism in the absence of corroberative biochemical proof. (*b*) Very high concentrations of some inhibitors are needed for them to be effective, due in part, perhaps, to their inability to penetrate the seeds. In those tissues that they do penetrate the possibility arises of "pharmacological" side-effects due to their high concentrations. (*c*) The specificity of RNA and protein synthesis inhibitors is not as narrow as was once assumed. Actinomycin D and cordycepin, sometimes assumed to be specific mRNA synthesis inhibitors, suppress synthesis of all types of RNA, and also have side effects. Some protein synthesis inhibitors reduce respiration. (*d*) To date, no evidence has been provided that when an inhibitor is effective it is specifically eliminating hormone-induced RNA or protein synthesis. This may be difficult to achieve, but the possibility must be considered that inhibitors can affect "basal" metabolism as well as, or instead of, hormone-induced metabolism. Reduction of basal metabolism, which is presumably essential for the general metabolic well-being of the seed, could result indirectly in reduced germination.

Several direct biochemical studies have been made on the effects of GA on protein and nucleic acid synthesis during germination. In hazel (*Corylus avellana*) embryonic axes there is an increase in total RNA 2-3 days after the start of imbibition of GA, compared to water controls (Jarvis *et al.*, 1968a). This increase is very small, however, prior to the observed increase in axis fresh weight which occurs 3-4 days after the start of imbibition and is the first physical indication of a GA effect (Fig. 1A). Axis elongation commences after 5 days. Synthesis of RNA increases within 16-24 hr of the start of imbibition of GA, and to a somewhat greater extent than in axes of water-imbibed seeds, though the largest increase is after the increase in fresh weight (Fig. 1B) (Jarvis *et al.*, 1968b). No RNA synthesis was detected in water- and GA-imbibed axes for some 12-24 hr after initial imbibition (Fig. 1B). In light of the current recognition that seed tissues commence RNA synthesis soon after the initiation of imbibition, the sensitivity of the techniques used here to detect RNA synthesis can be questioned.

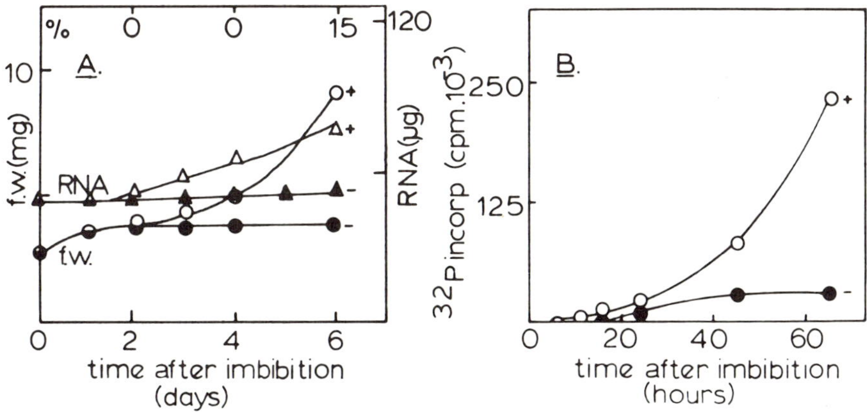

FIGURE 1. (A) Changes in fresh weight of embryonic axes of hazel in the presence (O+) and absence (●-) of GA_3, and the changes in RNA content in the presence (Δ+) and absence (▲-) of the hormone. Top line: percentage of seeds germinated at the times indicated (after Jarvis et al., 1968a). (B) The amount of $^{32}PO_4$ incorporated into RNA of embryonic axes of hazel in the presence (O+) or absence (●-) of GA_3 (after Jarvis et al., 1968b).

On the basis of these experiments, and the observation that GA-induced increases in DNA template availability and RNA polymerase activity may precede increases in RNA synthesis, Jarvis *et al.* (1968a) postulated Scheme 1 for the action of GA:

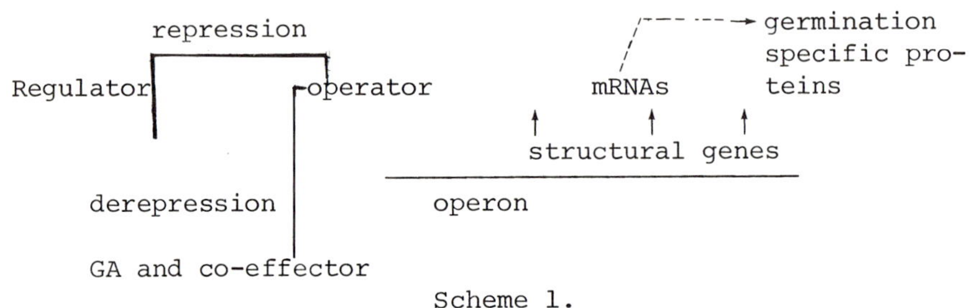

Scheme 1.

There is no strong evidence in favor of this, however. First of all, it has not been shown that such a mechanism of genetic control exists in plants. Furthermore, although increases in rRNA (Jarvis *et al.*, 1968a) and tRNA (Jarvis and Hunter, 1971) have been reported in GA-induced hazel axes, no synthesis of mRNA has been demonstrated (modern techniques involving extraction of poly(A)$^+$-rich RNA have not been tried). It can be argued that GA may not be enhancing germination-associated RNA or protein synthesis, but merely elevating the general level of synthetic metabolism, of which these are but one manifestation. Thus the evidence that GA controls hazel embryonic axis germination through synthesis of specific mRNAs and proteins is, at best, tentative.

The evidence in other seeds is hardly more convincing. GA enhances uridine incorporation into RNA, and leucine incorporation into proteins of wild oat (*Avena fatua*) embryos prior to radicle expansion (Chen and Park, 1973), implying that the hormone stimulates transcription and hence subsequent translation. But the above arguments apply here too.

On the other hand, in the wheat embryo there are apparently no effects of GA on RNA synthesis during the first 12 hr after imbibition starts (i.e., no enhanced transcription), but there is stimulation of protein synthesis (Fig. 2) (Chen and Osborne, 1970).

An important piece of information is missing from these results, for it was not stated when radicle elongation commences only that it had occurred by hour 12. Others have detected fresh weight increases (i.e., radicle growth) of isolated wheat embryos as early as 6-8 hr from initial imbibition (e.g., Marcus, 1969). Hence one cannot eliminate the possibility that the GA-induced increase in protein synthesis is associated with growth rather than with germination per se. There is another disturbing fea-

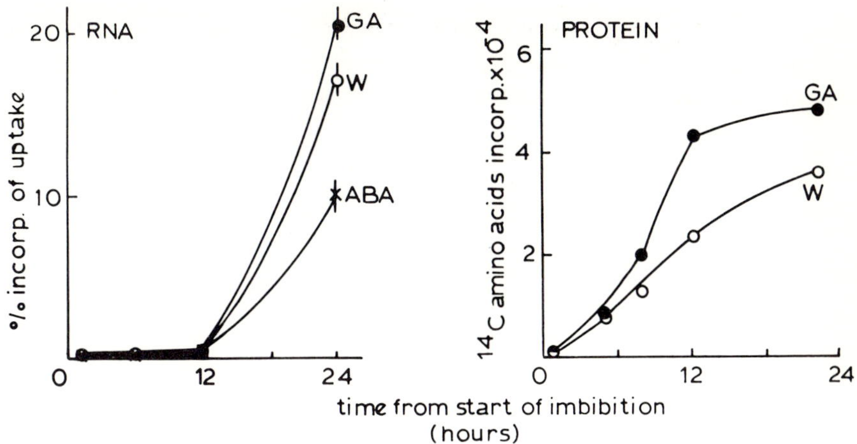

FIGURE 2. Effects of GA and ABA on the incorporation of ^3H-uridine into RNA and ^{14}C-amino acids into protein by isolated wheat embryos. w: water-imbibed controls (after Chen and Osborne, 1970).

ture of these results. Spiegel *et al.* (1975) have demonstrated that both ribosomal and messenger RNA are synthesized from the earliest times following the start of imbibition on water, yet Chen and Osborne found none for the first 12 hr (Fig. 2).

Maize embryos imbibed in GA synthesize 20-40% more RNA over the first 4 hr than do water controls (Wielgat *et al.*, 1974). Interestingly, though, it is the scutellum that responds most positively to GA, with a several-fold increase in ^{32}P-orthophosphate incorporation into two rRNA species. Cereal embryos, when isolated, usually include the scutellum as an integral part. Yet this, unlike the axis, is a nongrowing tissue. Thus, when studying GA-induced effects on isolated cereal embryos, care should be taken to localize the changes in synthesis; for changes in metabolism of the scutellum may be unrelated to germination, which takes place (chiefly) in the axis.

In lettuce, GA stimulates polyribosome and protein synthesis (Fountain and Bewley, 1976), but not RNA synthesis (Fountain, 1974) above water control levels, and prior to completion of germination. Similarly, charlock (*Sinapis arvensis*) shows increases in amino acid and protein synthesis in the presence of GA (Edwards, 1976), and autoradiography has shown that much of the synthesis is confined to the apical meristems. In either seed it is not known if the proteins that are synthesized are linked to germination. Prevention of GA-induced lettuce seed germination by ABA is accompanied by a partial reversal of GA-stimulated protein synthesis (Fig. 3). It may be argued that

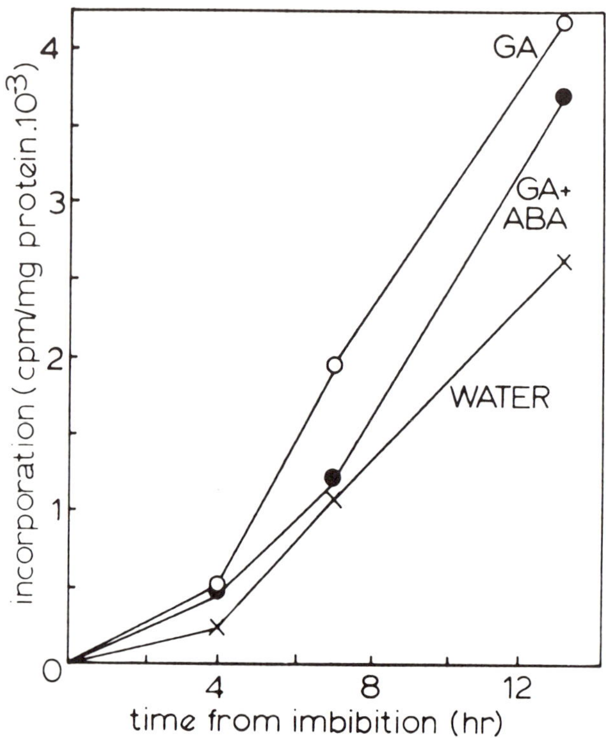

FIGURE 3. The effects of GA_3 (0.29 mM) and GA_3 + ABA (0.04 mM) on the incorporation of ^3H-leucine into protein by intact lettuce seeds (based on Fountain and Bewley, 1976; and Fountain, 1974).

ABA is inhibiting germination by reversing GA-induced protein synthesis which is essential for dormancy breaking. On the other hand, the possibility that ABA is simply reducing basal protein synthesis and indirectly affecting seed germinability cannot be ignored. Indeed, studies on isolated lettuce embryos have shown that ABA suppresses polyribosome formation and protein synthesis below those levels found in water-imbibed control embryos (Fountain and Bewley, 1976).

Inhibition of germination of a number of seed species by ABA is accompanied by a general reduction in nucleic acid synthesis, for example, in wheat (Fig. 2), lettuce (Fountain, 1974), pear (Khan and Heit, 1968) and ash (Villiers, 1968) embryos, and in germinating bean axes (Walbot *et al.*, 1975), but there is no substantive evidence that any one species of RNA is inhibited in preference to any other. There are claims that ABA is without effect on either nucleic acid or protein synthesis in some seeds and seed axes; invariably the experimental technique upon which these claims are based is suspect.

Cytokinins stimulate the germination of very few seed species, though they are known to reverse the inhibitory effects of ABA. In bean axes (Sussex *et al.*, 1975) and pear embryos (Khan and Heit, 1968) this reversal is accompanied by an increase in RNA synthesis, and relief of ABA-induced "dormancy" in lettuce embryos by benzyladenine is accompanied by an elevation in polyribosome levels and protein synthesis (Fig. 4).

Perhaps, one manifestation of ABA action in preventing seed germination is an inhibition of RNA and protein synthesis which is essential for radicles to commence elongation, though it could be inhibiting the RNA and protein synthesis component of basal metabolism. Cytokinins somehow prevent, or reverse, this inhibition.

In brief summary: GA does appear to promote RNA and/or protein synthesis in some seeds, and ABA is inhibitory. This latter inhibition can be reversed by cytokinin. As yet, there

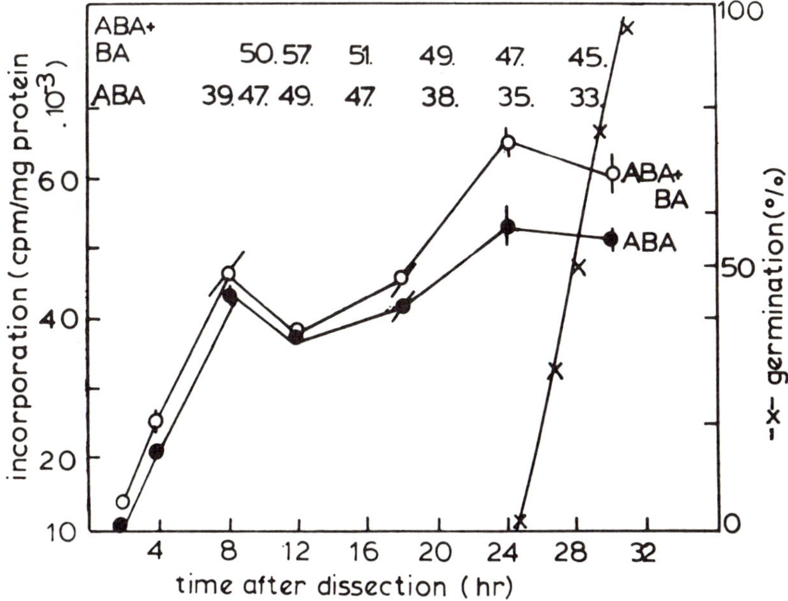

FIGURE 4. The effects of ABA (0.02 mM) and ABA + 6-benzyladine (0.01 mM) on ^{3}H-leucine incorporation into protein by isolated lettuce embryos. Time course of germination of embryos on ABA + BA shown as X——X. Percentage of polyribosomes at various times of embryos on ABA + BA or ABA alone are shown at the top of the figure (based on Fountain and Bewley, 1976 and Fountain, 1974).

is no compelling evidence that any of these hormones acts at the level of transcription and translation to regulate the synthesis of proteins essential for the germination process--if, indeed, such proteins exist.

EFFECTS OF HORMONES ON OTHER CHEMICALS, ON RESPIRATION AND RESPIRATORY PATHWAYS

Little research has been done on the effects of hormones in oxygen comsumption by seeds, and even less on their effects on ATP synthesis. Excised embryos of wild oat whose dormancy has been broken by GA do not consume significantly greater quantities of oxygen prior to completion of germination than do dormant controls imbibed in water (Simmonds and Simpson, 1971). Lettuce seeds maintained in the dormant state on water do not consume significantly less oxygen than seeds imbibed on GA for up to 8 hr after imbibition (Fig. 5). Radicle protrusion from the first seeds of the population occurs at about 10 hr. The addition of ABA to GA prevents germination, but does not reduce oxygen consumption. It is not possible to conclude that enhanced or suppressed respiration are important facets of dormancy-making or breaking.

Preliminary results on ATP synthesis in lettuce seeds indicate that over the first 12 hr from the start of imbibition endogenous levels of ATP are not elevated by GA, nor are they suppressed by ABA, compared to water controls (Krochko, unpublished data). In germinating bean axes, ABA depletes the ATP pool by about 10% at concentrations that inhibit growth. But this is probably unimportant, for the addition of BA can completely reverse the growth inhibition from ABA without appreciably elevating the size of the ATP pool (Sussex *et al.*, 1975).

The major thrust in research on respiration in relation to dormancy breaking has not focused on the action of the known plant hormones but on the effects of other chemicals that stimulate germination. Germination of a number of dormant seeds has been achieved by imbibing them at elevated oxygen tensions (Roberts and Smith, 1977). It is perhaps surprising, therefore, that certain respiratory inhibitors that reduce oxygen consumption also are effective dormancy breaking agents. These include (*a*) hydrogen acceptors (nitrite, nitrate, methylene blue); (*b*) terminal oxidase inhibitors (cyanide, azide, hydroxylamine); (*c*) citric acid cycle inhibitors (monofluoracetate, malonate); (*d*) glycolysis inhibitors (iodoacetate, sodium fluoride); and (*e*) sulfur-containing compounds (thiourea, mercaptoethanol, dithiothreitol). In an attempt to explain these effects Roberts has proposed a mechanism for dormancy breaking that depends on

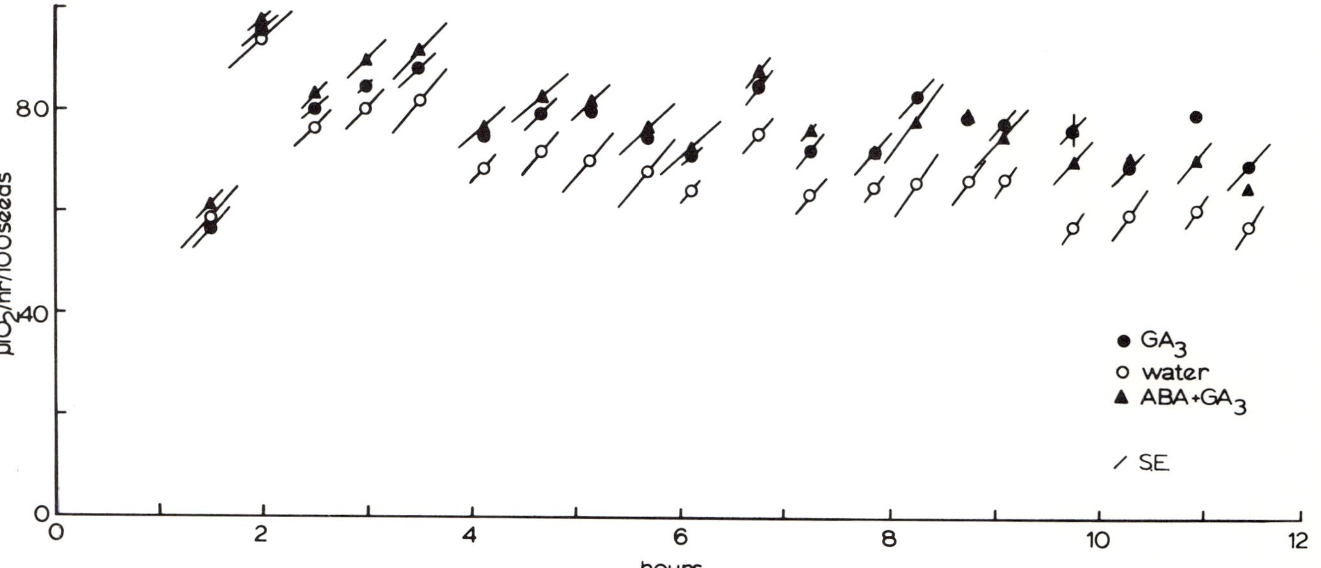

FIGURE 5. Oxygen consumption by lettuce seeds maintained in darkness at 25°C on water, 100 μg/ml GA_3, or 100 μg/ml GA_3 + 0.5 mM ABA. About 10% of the seeds had germinated after 12 hr on GA_3 (Krochko and Bewley, unpublished data).

shifts in respiratory metabolism (Roberts 1969, 1973; Roberts and Smith, 1977). His initial premise is that dormant seeds have certain respiratory deficiencies that are absent from non-dormant ones. As shown in Fig. 6, one such deficiency is imposed by the citric acid cycle, which is highly active and utilizes available oxygen to the exclusion of other oxygen-requiring processes. It is these latter processes that must operate in order for dormancy to be broken. Thus raising the ambient oxygen concentration provides an excess of the gas over and above the requirement of the citric acid cycle, an excess which can be used by the normally oxygen-starved dormancy-breaking processes. Roberts' proposal also assumes that attenuation of the electron transport chain by inhibitors of the citric acid cycle and terminal oxidation reactions, etc., reduces their requirement for oxygen, thus freeing more to allow the dormancy-breaking processes to proceed. Net oxygen consumption by dormant and nondormant seeds--including those whose dormancy has been broken by both chemical and natural (e.g., after-ripening) agents may be expected to be the same, only the pathway through which it is utilized will be different.

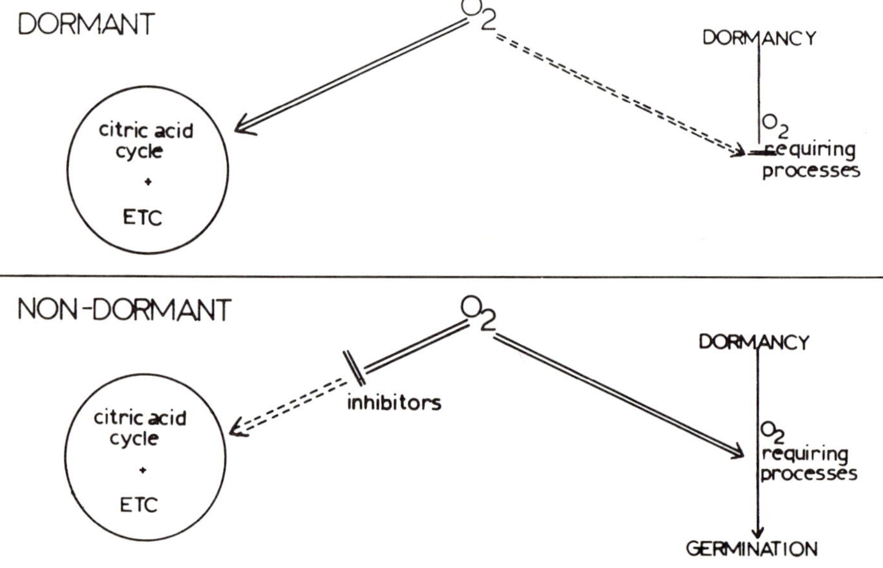

FIGURE 6. The role of the citric acid cycle in dormant or nondormant seeds.

The pentose phosphate pathway has been proposed as the oxygen-requiring process whose operation is essential for dormancy-breaking (Roberts 1969, 1973). This pathway is cyanide insensitive, and requires reoxidation of NADPH for its operation. This is claimed to be the rate-limiting reaction under conditions of low oxygen availability. Promotion of dormant seeds by hydrogen acceptors is assumed to be mediated through reoxidation of NADPH, thus stimulating operation of the pentose phosphate pathway.

Approximate estimations of the relative activities of the glycolytic and pentose phosphate pathways in seeds have been obtained by comparing their abilities to respire glucose-6-^{14}C and glucose-1-^{14}C. For details of the technique see Bewley and Black (1978, p. 151). Essentially it involves collection and estimation of the $^{14}CO_2$ released from both position 1 and 6 of the glucose molecule, the results being expressed as a C_6/C_1 ratio. A high ratio indicates an active glycolytic pathway; the lower the ratio, the more active is the pentose phosphate pathway. Nondormant grains of barley (cv. Pallas) show a lower C_6/C_1 ratio than do dormant ones (Roberts, 1969), indicating a more active pentose phosphate pathway in those grains that are going to germinate. Release from dormancy of wild oat and barley grains by chemical agents is associated with C_6/C_1 ratio changes prior to radicle elongation (Table I), though in some cases the changes are very small.

The validity of using C_6/C_1 ratios to obtain quantitative estimations of the activities of the two pathways is subject to some doubt (Davies *et al.*, 1964), but even accepting that the pentose phosphate pathway is somehow important for dormancy breaking requires that two important questions be answered. First, how is the pentose phosphate pathway blocked in dormant seeds? Second, what does the pathway provide that is essential for stimulating germination?

In an attempt to answer the first question, direct comparisons have been made between the extractable activities of glucose-6-phosphate dehydrogenase (G6PdH) and 6-phosphogluconate dehydrogenase (6PGdH)--the first two enzymes of the pentose phosphate pathway--in dormant and nondormant grains. In nondormant wild oat grains the levels of G6PdH rise with increasing time after imbibition starts, but decline in dormant grains; the same is true for 6PGdH, but in a less dramatic fashion (Fig. 7). Activities of enzymes of the glycolytic pathway either remain constant or rise equally in dormant and nondormant grains. It has been suggested, therefore, that the ability to maintain, or increase, initial levels of pentose phosphate pathway enzymes is essential for germination (Kovacs and Simpson, 1976). The evidence is not compelling, however. For radicle emergence from the nondormant grains commences after about 12 hr, about 50% of the population has germinated by 16 hr,

TABLE I. Changes in the C_6/C_1 Ratio Prior to Radicle Emergence Associated with Dormancy Breaking of Cereals by Various Chemicals

Species	Chemical treatment	C_6/C_1 ratio[a]		Reference
		Treated	Untreated dormant control	
Avena fatua *(embryos)*	*GA (50ppm)*	*0.64 - 0.65*	*0.76 - 0.84*	*Simmonds & Simpson (1971)*
	Malonate (10^{-2}M)	*0.73 - 0.80 (83)*	*0.83 - 0.85 (17)*	*Simmonds & Simpson (1972)*
Barley grains cv. Golden Promise	*2-Mercaptoethanol (5×10^{-2}M)*	*0.03 - 0.04 (89)*	*0.36 - 0.42 (8)*	*Roberts (1973)*
	Dithiothreitol (10^{-1}M)	*0.07 - 0.09 (90)*	*0.2 - 0.37 (10)*	*Roberts (1973)*
Barley grains cv. Proctor	*GA (10^{-3}M)*	*0.11 (100)*	*0.24 (10)*	*Roberts (1969)*
	$NaNO_2$ (10^{-2}M)	*0.20 (32)*	*0.22 (10)*	*Roberts (1969)*
	KCN (10^{-3}M)	*0.16 (26)*	*0.19 (10)*	*Roberts (1969)*

[a]*Parentheses after the C_6/C_1 ratios for treated and untreated dormant control values show, where known, the final germination percentage promoted by chemical treatment.*

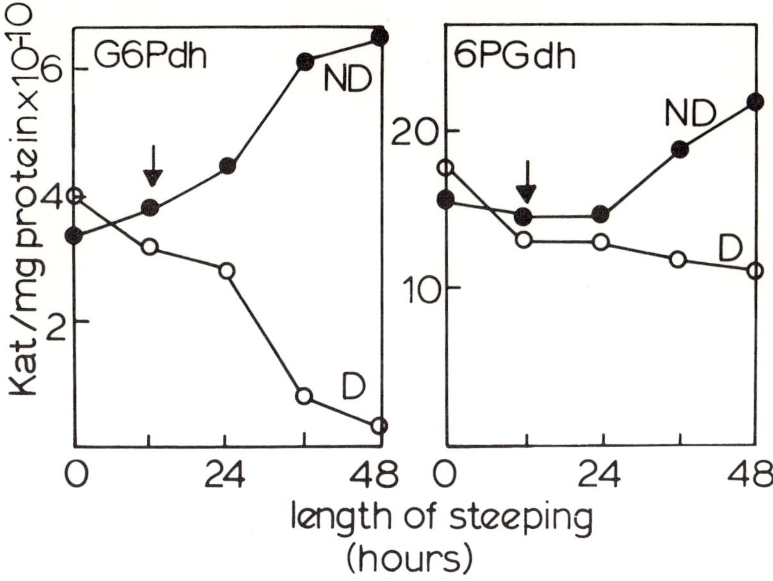

FIGURE 7. Glucose-6-phosphate dehydrogenase (G6PdH) and 6-phosphogluconate dehydrogenase (6PGdH) activities in imbibed dormant (D) and nondormant (ND) grains of wild oat. Arrows indicate time of first radicle emergence (after Kovacs and Simpson, 1976).

and all by 24 hr (Simpson, personal communication). Hence it is difficult to ascribe dormancy to differences in activities of these pentose phosphate pathway enzymes alone, for their levels in dormant and nondormant grains are similar over the first 12 hr, and yet only the latter will subsequently germinate. In barley, there are no differences in activities of these enzymes between dormant and nondormant grains prior to the completed germination of the latter (Roberts, 1973); nor are there any differences in the levels of NAD/NADH and NADP/NADPH between dormant and nondormant grains.

As mentioned above, it has been suggested that the rate of reoxidation of NADPH is a limiting reaction in dormant grains. This could result from oxygen starvation or from the relative inactivity of an NADPH oxidase system in dormant grains (although none are known that appear to fit the requirements of the hypothesis). Related to this is an idea put forward by Hendricks and Taylorson (1974, 1975) to explain the promotive action of thiourea, nitrite, and hydroxylamine salts on *Amaranthus alba* seeds and of thiourea on lettuce seeds. Dormancy breaking by these chemicals is accompanied by an irreversible inhibition of catalase activity, which is most striking for thiourea-induced lettuce seed germination (Fig. 8). Neither respiration nor peroxidase activity are affected by these chemicals at concentrations that promote germination. These observations have led to the suggestion that the pathway operates as in Scheme 2. Thus in dormant seeds, catalase removes H_2O_2 from a pathway whereby oxidizing power is passed through quinones to NADPH, allowing for its oxidation and hence for operation of the pentose phosphate pathway. Inhibition of catalase activity spares H_2O_2 for peroxidase action, which results in eventual stimulation of the pentose phosphate pathway.

There are weaknesses in this proposal, however: (*a*) It has not been demonstrated that these or other enzymes are cooperatively linked together in the manner shown; (*b*) Roberts and Smith (1977) claim that there are no shifts in C_6/C_1 ratios when lettuce seed dormancy is broken by light, GA, or mercaptoethanol; (*c*) that NADPH reoxidation is essential for dormancy breaking is only an assumption; (*d*) the source of

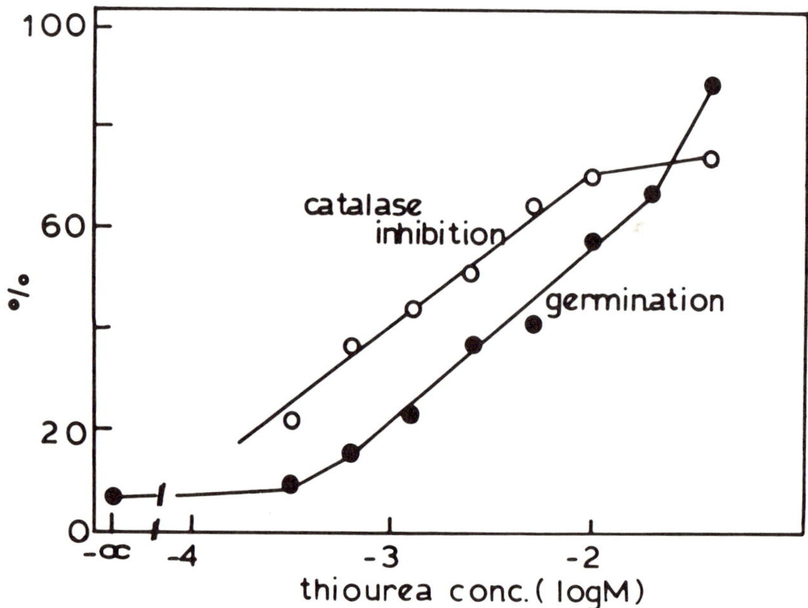

FIGURE 8. The effects of various concentrations of thiourea on germination of Grand Rapids lettuce seeds at 25°C (counted after 48 hr from the start of imbibition) and on the inhibition of extracted catalase activities (measured 24 hr from the start of imbibition) (after Hendricks and Taylorson, 1975).

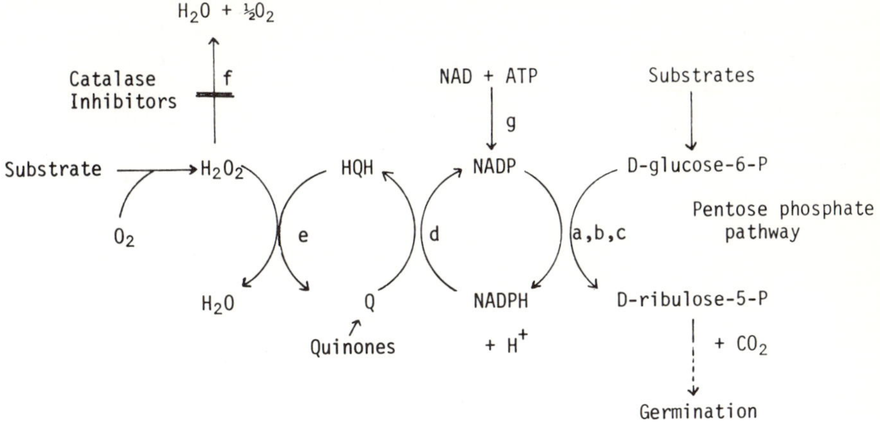

Scheme 2.

Enzymes: (a) D-glucose-6-phosphate: NADP 1-oxidoreductase; (b) 6-phospho-D-gluconate-5-lactone hydrolase; (c) 6-phospho-D-gluconate: NADP 2-oxidoreductase (decarboxylating); (d) NAD (P)H quinone oxidoreductase; (e) peroxidase; (f) catalase; (g) ATP: NAD 2'-phosphotransferase.

H_2O_2 for endogenous catalase activity prior to visible germination is not clear. It is hardly likely to come from β-oxidation of fatty acids in the glyoxysomes because mobilization of the major lipid reserves in lettuce seeds does not occur until many hours after radicle emergence (Halmer *et al.*, 1978).

Another weakness in the proposal relates to the time at which catalase activity was measured. As noted in the legend to Fig. 8 catalase was only assayed 24 hr from the start of imbibition, and it was not stated how many seeds had completed germination at this time. Experiments carried out in my laboratory on the same cultivar of lettuce (Grand Rapids) show that at 10^{-1} and 10^{-2} M thiourea a considerable number of seeds have already germinated after 24 hr (Fig. 9), and hence one might argue that thiourea could be affecting catalase activity associated with growth rather than germination per se. Thus to determine whether or not suppression of catalase activity is associated with germination, catalase assays were conducted at times prior to radicle protrusion (Table II).

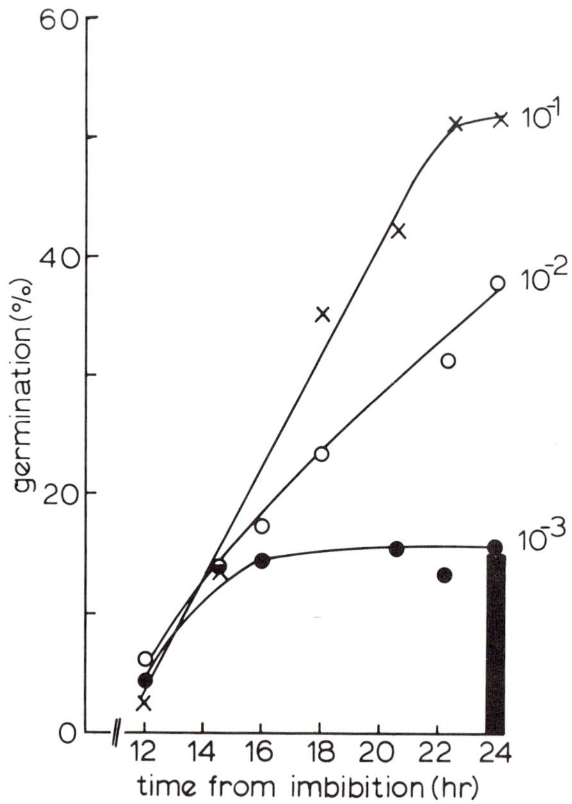

FIGURE 9. Time course of germination of Grand Rapids lettuce seeds imbibed at 25°C on different concentrations of thiourea or water (dark bar graph).

Catalase activity in water-imbibed seeds remains fairly constant until after about 12 hr when germination is completed for some 5% of the population. These germinated seeds do not account for the increase in enzyme activity, however, for even after removing them catalase activity of the ungerminated population is higher than after 9 hr (results not presented). In the presence of 10^{-2} M thiourea, catalase activity is suppressed at all times during germination, but with 10^{-1} M thiourea it is not, at least until the twelfth hour. Why the lower concentration suppresses enzyme activity more than the higher one is not readily apparent, though the higher concentration promotes considerably more germination. After 24 hr, as reported by Hendricks and Taylorson (1975), thiourea at both concentrations considerably reduces catalase activity. On the basis of the

TABLE II. The Effects of Thiourea on Catalase Activity in Lettuce Seeds during Germination

	Time from start of imbibition (hr)				
	3	6	9	12	24
	Catalase activity (units/250 mg dry weight)				
Treatment					
Thiourea 10^{-1} M	272 ± 4	244 ± 10	236 ± 5	238 ± 3	191 ± 16
Thiourea 10^{-2} M	173	203	191 ± 2	224	228 ± 17
Water	215 ± 8	227 ± 6	236 ± 16	291 ± 2	416 ± 9
Activity in dry seeds: 203.6 units/250 mg dry weight					

(Bewley and Matowe, unpublished data).

experiment outlined in Table II it is not easy to argue that thiourea-induced germination is mediated through suppression of catalase activity, though it is not possible to argue completely against it. There appears to be no relationship between concentrations of thiourea, their germination promotion activities, and the amount of catalase activity suppression except, perhaps, after 12 hr. Even then the suppression of activity is only about 20% at both concentrations (despite the same degree of suppression, a different final level of germination is achieved). It is of interest to note that GA- and red-light-induced germination results in a slight increase in catalase activity at 12 hr (though some seeds have germinated at this time) and no suppression prior to visible germination (results not presented).

In summary, then, the suggestion that the pentose phosphate pathway is involved in dormancy breaking has some indirect support, but how and why it is involved remains to be shown. The action of various chemicals in promoting germination of dormant seeds has been noted, and while it is of both importance and interest to determine their action at the molecular level, only limited progress has been made so far. Some thought-provoking ideas have been proffered, but evidence to support them is moot.

SUMMARY

Several studies have attempted to elucidate the molecular action of hormones and other chemicals in breaking dormancy. Some have purported to show increased transcription and/or translation in hormone-treated seeds compared with dormant controls. For example, gibberellin appears to promote RNA and/or protein synthesis, and ABA inhibits this process; this inhibition, in turn, can be reversed by CK. That these hormones promote or inhibit synthesis of mRNA and/or proteins that are related specifically to germination has not been demonstrated.

Germination can be promoted by a number of chemicals whose actions are thought to limit or to modify respiratory metabolism. It has been proposed that activation of the pentose phosphate pathway is an important facet of dormancy breaking. The evidence is reviewed critically, and the conclusion is drawn that it lacks substantial support.

REFERENCES

Bewley, J. D., and Black, M. (1978). "Physiology and Biochemistry of Seeds in Relation to Germination. 1. Development, Germination and Growth." Springer-Verlag, Berlin.

Black, M., and Bewley, J. D. (1979). "Physiology and Biochemistry of Seeds in Relation to Germination. 2. Viability, Dormancy and Environmental Control." Springer-Verlag, Berlin (in press).

Chen, D., and Osborne, D. J. (1970). Hormones in the translational control of early germination in wheat embryos. *Nature 226,* 1157-1160.

Chen, S. C. C., and Park, W. M. (1973). Early actions of gibberellic acid on the embryo and on the endosperm of *Avena fatua* seeds. *Plant Physiol. 52,* 174-176.

Davies, D. D., Giovanelli, J., and Ap Rees, T. (1964). "Plant Biochemistry," pp. 85-155. Blackwell Scientific Publications, Oxford.

Edwards, M. (1976). Dormancy in seeds of charlock (*Sinapis arvensis* L.): Early effects of gibberellic acid on the synthesis of amino acids and proteins. *Plant Physiol. 58,* 626-630.

Fountain, D. W. (1974). On the physiology and biochemistry of dormancy and germination in light-sensitive lettuce seeds. PhD. thesis. University of Calgary.

Fountain, D. W., and Bewley, J. D. (1973). Polyribosome formation and protein synthesis in imbibed but dormant lettuce seeds. *Plant Physiol. 52,* 604-607.

Fountain, D. W., and Bewley, J. D. (1976). Lettuce seed germination: Modulation of pregermination protein synthesis by gibberellic acid, abscisic acid, and cytokinin. *Plant Physiol. 58,* 530-536.

Frankland, B., Jarvis, B. C., and Cherry, J. H. (1971). RNA synthesis and the germination of light-sensitive lettuce seeds. *Planta (Berl.) 97,* 39-49.

Halmer, P., Bewley, J. D., and Thorpe, T. A. (1978). Degradation of the endosperm cell walls of *Lactuca sativa* L., cv. Grand Rapids. Timing of mobilization of soluble sugars, lipid and phytate. *Planta (Berl.) 139,* 1-8.

Hendricks, S. B., and Taylorson, R. B. (1974). Promotion of seed germination by nitrate, nitrite, hydroxylamine, and ammonium salts. *Plant Physiol. 54,* 304-309.

Hendricks, S. B., and Taylorson, R. B. (1975). Breaking of seed dormancy by catalase inhibition. *Proc. Nat. Acad. Sci. (Wash.) 72,* 306-309.

Jarvis, B. C., and Hunter, C. (1971). Changes in the capacity for protein synthesis in embryonic axes of hazel fruits during the breaking of dormancy by GA_3. *Planta (Berl.) 101*, 174-179.

Jarvis, B. C., Frankland, B., and Cherry, J. H. (1968a). Increased nucleic-acid synthesis in relation to the breaking of dormancy of hazel seed by gibberellic acid. *Planta (Berl.) 83*, 257-266.

Jarvis, B. C., Frankland, B., and Cherry, J. H. (1968b). Increased DNA template and RNA polymerase associated with the breaking of seed dormancy. *Plant Physiol. 43*, 1734-1736.

Khan, A. A., and Heit, C. E. (1968). Selective effect of hormones on nucleic acid metabolism during germination of pear embryos. *Biochem. J. 113*, 707-712.

Kovacs, M. I. P., and Simpson, G. M. (1976). Dormancy and enzyme levels in seeds of wild oats. *Phytochem. 15*, 455-458.

Marcus, A. (1969). Seed germination and the capacity for protein synthesis. *In* "Dormancy and Survival" (H. W. Woolhouse, ed.). Cambridge Univ. Press. *Symp. Soc. Exp. Biol. 23*, 143-160.

Roberts, E. H. (1969). Seed dormancy and oxidation processes. *In* "Dormancy and Survival" (H. W. Woolhouse, ed.). Cambridge Univ. Press. *Symp. Soc. Exp. Biol. 23*, 161-192.

Roberts, E. H. (1973). Oxidative processes and the control of seed germination. *In* "Seed Ecology" (W. Heydecker, ed.), pp. 189-218. Butterworths, London.

Roberts, E. H., and Smith, R. D. (1977). Dormancy and the pentose phosphate pathway. *In* "The Physiology and Biochemistry of Seed Dormancy and Germination" (A. A. Khan, ed.), pp. 385-411. Elsevier/North-Holland Biomedical Press, Amsterdam.

Simmonds, J. A., and Simpson, G. M. (1971). Increased participation of pentose phosphate pathway in response to after ripening and gibberellic acid treatment in caryopses of *Avena fatua*. *Can. J. Bot. 49*, 1833-1840.

Simmonds, J. A., and Simpson, G. M. (1972). Regulation of the Krebs cycle and pentose phosphate pathway activities in the control of dormancy of *Avena fatua*. *Can. J. Bot. 50*, 1041-1048.

Spiegel, S., Obendorf, R. L., and Marcus, A. (1975). Transcription of ribosomal and messenger RNAs in early wheat embryo germination. *Plant Physiol. 56*, 502-507.

Sussex, I., Clutter, M., and Walbot, V. (1975). Benzyladenine reversal of abscisic acid inhibition of growth and RNA synthesis in germinating bean axes. *Plant Physiol. 56*, 575-578.

Villiers, T. A. (1968). An autoradiographic study of the effect of the plant hormone abscisic acid on nucleic acid and protein metabolism. *Planta (Berl.) 82*, 342-354.

Walbot, V., Clutter, M., and Sussex, I. (1975). Effects of abscisic acid on growth, RNA metabolism, and respiration in germinating bean axes. *Plant Physiol. 56*, 570-574.

Wielgat, B., Wasilewska, L. D., and Kleczkowski, K. (1974). RNA synthesis in germinating maize seeds under hormonal control. *In* "Plant Growth Substances 1973," pp. 593-598. Hirokawa Publishing Co., Tokyo.

Woodstock, L. W., and Toole, V. K. (1976). Respiration of Grand Rapids lettuce (*Lactuca sativa* L.) seeds in relation to chemical and photocontrol of germination. *Plant Cell Physiol. 18*, 1-8.

HORMONAL CONTROL OF ENDOSPERM FUNCTION DURING GERMINATION

J. V. Jacobsen
T. J. V. Higgins
and
J. A. Zwar

Division of Plant Industry
C S I R O
Canberra, ACT 2601
Australia

During the growth cycle of seed plants, the vegetative body absorbs raw materials from the environment, processes them where necessary, and deposits a fraction of the total in the storage tissues of its seeds. Seed reserves are both inorganic and organic, the latter consisting commonly of lipids, carbohydrates, proteins, oxalate, and phytate, and they are deposited in various seed tissues including perisperm, endosperm and in the axis of the embryo as well as its cotyledons. The reserves are commonly deposited in specialized organelles such as amyloplasts, protein bodies, aleurone grains, and spherosomes or lipid droplets.

During seed germination, the storage tissues become the source of nutrients for the enlarging and developing embryo. The endosperm and perisperm reserves break down, the tissues are consumed by the enlarging embryo, and the cotyledonary reserves are mobilized into the axis. The biochemistry of some of these processes, for example, protein and lipid degradation, has received considerable attention (Ching, 1972; Millerd, 1975; Ashton, 1976), but the factors that regulate these processes are relatively unknown, although plant hormones have been considered to be prime candidates. There is little convincing evidence for axial control and for hormonal involvement in mobilization of cotyledonary reserves; however, the evidence for axial control and participation of hormones in

ISBN 0-12-602050-7

endosperm physiology is, at least in some cases, considerably more convincing. It is the purpose of this paper to review aspects of endosperm and its function.

Because it not a widely understood tissue, endosperm will be considered in a broad sense. The structure, composition, and physiology of the cells will be considered first and followed by a discussion of hormonal control of enzyme induction at the nucleic acid and protein level.

ENDOSPERM STRUCTURE AND COMPOSITION

Information on the structure of mature endosperm is difficult to obtain since the subject has only occasionally attracted the attention of morphologists and anatomists. Indeed, the bulk of the present knowledge is contained in two major studies completed many years ago (Netolitzky, 1926; Martin, 1946). These studies are primarily zonal in approach and there is little cellular information, but several basic facts emerge: (*a*) Endosperm occurs to some degree in mature seeds of almost all families and may vary in amount from occupying the bulk of the seed volume to a few layers of cells. (*b*) In most plant families, mature endosperm appears to consist of living cells, although in some cases, as in the Gramineae and in some Leguminosae, some of the endosperm is dead at maturity. (*c*) Living endosperm cells commonly contain large numbers of aleurone grains; they store large quantities of lipid, and they often have thickened cell walls. (*d*) The endosperm can be lateral to the embryo or it can completely surround the embryo, in which case the embryo must penetrate the endosperm before germination (radicle emergence) can occur.

There have been few studies of living endosperm cells using modern light and electron microscopic techniques, but the available data confirm that there is a degree of uniformity in their structure and apparent composition. Figure 1 compares electron micrographs of mature endosperm cells of celery (*Apium graveolens* L.) and lettuce (*Lactuca sativa* L.) and aleurone cells of barley (*Hordeum vulgare* L.). In all of these cells, the most abundant organelles are lipid droplets (or spherosomes) and aleurone grains. In castor bean (Vigil, 1970) and celery, the cells are packed with these organelles, and other cytoplasmic organelles such as mitochondria are rare and undeveloped (Mollenhauer and Totten, 1970; Vigil, 1970; Jacobsen *et al.*, 1976), although in lettuce and barley cells the lipid droplets and aleurone grains are less abundant and mitochondria, microbodies, endoplasmic reticulum, dictyosomes, and a variety of vesicles are more evident (Paleg and Hyde, 1964; Eb and Niewdorp, 1967; Jones, 1969a,b; 1974). The aleurone grains in

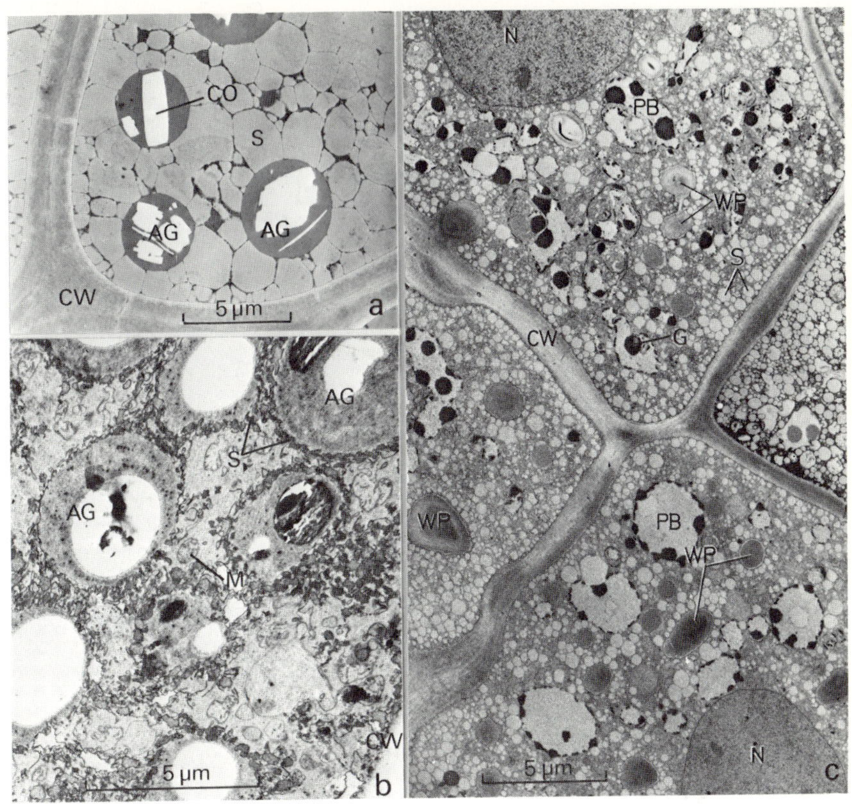

FIGURE 1. Electron micrographs of endosperm cells of (a) celery, (b) barley (aleurone), (c) lettuce. S, spherosome; CW, cell wall; M, mitochondrion; G, globoid; N, nucleus; AG, aleurone grain, same as PB, protein body; WP, wall protuberances; CO calcium oxalate crystal removed during secioning ((a) Jacobsen et al., *1976; (b) Jones, 1969a; (c) Jones, 1974; courtesy of Springer-Verlag).*

these cells are always large and usually contain one or more globoids containing potassium and magnesium phytate, probably complexed with protein. They may or may not contain a second inclusion termed, among other things, the protein-carbohydrate body in barley (Jacobsen *et al.*, 1971), the crystalloid in castor bean (Tully and Beevers, 1976; Youle and Huang, 1976), and inclusion type II in wheat (Fulcher, 1972). The cell walls vary from thin (0.5-1.0 µm) in castor bean to thick (1-5 µm) in barley, celery, and lettuce. Endosperm cells of some legumes such as fenugreek (*Trigonella foenum-graecum* L.) (Reid and Meier, 1972), carob (*Ceratonia siliqua*) (Seiler, 1977) and others (Corner, 1951) seem to have similar cytoplasmic organelles, although wall thickening (galactomannan deposition) in legumes like carob occurs to the extent that at maturity the lumen of the cell may be completely occluded. However, the fine structure of the cytoplasm of these cells has not been examined in detail.

STRUCTURAL CHANGES OCCURRING IN ENDOSPERM CELLS DURING GERMINATION

A review of the literature shows that living endosperm cells of different species undergo similar structural modifications during germination and in some cases following hormone treatment. During germination of castor bean (Vigil, 1970; Mollenhauer and Totten, 1970), fenugreek (Reid and Meier, 1972), carob (Seiler, 1977), wheat (Laidman *et al.*, 1974), and barley (Paleg and Hyde, 1964; Jones, 1969b,c; Jones and Price, 1970; Taiz and Jones, 1970) and during GA_3 treatment of isolated wheat and barley aleurone, changes occur in the endosperm cells which reflect lipid breakdown, gluconeogenesis, enzyme synthesis, and finally cell autolysis. The lipid bodies become fewer in number, mitochondria and glyoxysomes usually (castor bean and cereals) but not always (fenugreek) become more numerous, and plastids, ribosomes, rough endoplasmic reticulum, dictyosomes, and occasionally starch grains (castor bean, fenugreek) appear. The aleurone grains swell and appear to vacuolate. As time progresses the cells contain fewer spherosomes and more glyoxysomes and mitochondria; the aleurone grains lose their contents, swell more, and coalesce to form large vacuoles. Ultimately the cellular organelles appear to be consumed by autophagic vacuoles, although this process has been described in some detail only in the case of castor bean (Vigil, 1970). Some of the cytoplasmic changes occurring in barley aleurone are illustrated in Fig. 2.

The intracellular changes are usually associated with degradation of the cell walls. This process has received considerable attention in barley aleurone where the wall, which consists largely of pentosans (McNeil *et al.*, 1975), is probably broken down by pentosanases originating in the aleurone cells (Taiz and Honigman, 1976) leaving a thin layer of resistant wall of unknown composition adjacent to the cytoplasm (Taiz and Jones, 1970; Ashford and Jacobsen, 1974a,b) (Fig. 2) which presumably holds the cell together. Similar changes in cell walls as well as the cytoplasmic changes seem to occur in the band of cells adjacent to the enlarging embryo (Fig. 3b) in celery seed during germination. The aleurone grains swell (Fig. 3d) and the walls are degraded. In gibberellin-treated

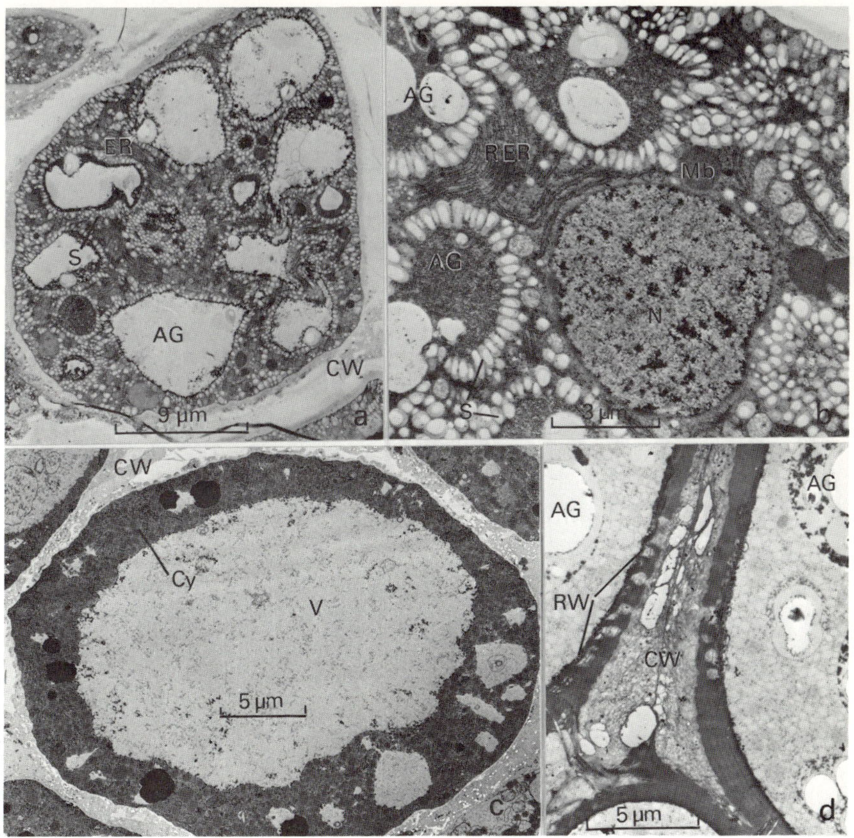

FIGURE 2. Electron micrographs showing changes occurring in barley aleurone cells during incubation with GA_3 (1.4 x 10^{-6} M) for (a) and (b) 10 hr, (c) 36 hr, and (d) 36 hr. AG, aleurone grain; S, spherosome; N, nucleus; Mb, microbody; M, mitochondrion; RER, rough endosplasmic reticulum; V, vacuole; CW, cell wall; Cy, cytoplasm; RW, residual wall (micrographs (a,b) Jones, 1969b; (c) Jones and Price, 1970; (d) N. A. Pyliotis, unpublished data; courtesy of Springer-Verlag).

deembryonated seeds, the endosperm undergoes similar changes and the cells separate from each other (Fig. 3e) (Jacobsen *et al.*, 1976). Figure 3f shows isolated endosperm cells that have been plasmolyzed to show the residual cell wall layer.

Taking into account the similarities in structure and in changes occurring during germination, there seems to be some cause for thinking that there is a common living endosperm cell type and that mature endosperm from genus to genus may

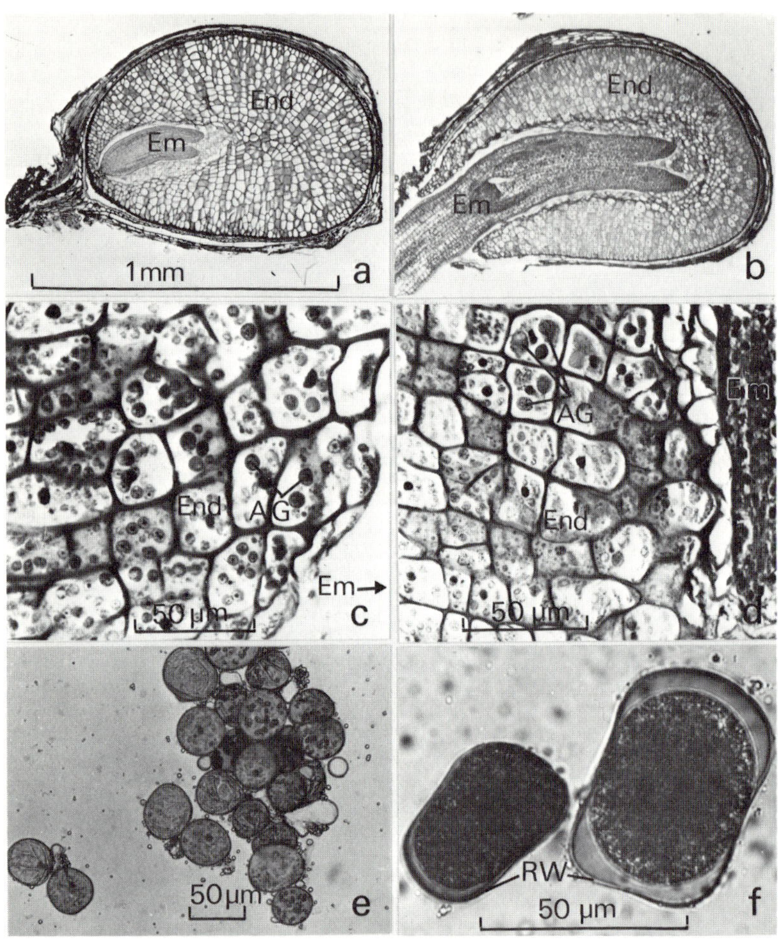

FIGURE 3. Light micrographs of celery seed showing changes occurring in endosperm during germination. (a) dry seed, (b) on water for 7 days, (c) cells adjacent to the embryo in dry seed, (d) cells adjacent to the embryo in 7 day imbibed seed, (e) separate whole cells obtained from isolated endosperm incubated with GA_3 for 6 days, (f) cells shown in (e) were plasmolyzed with mannitol to show the residual wall (RW). End, endosperm; Em, embryo; AG, aleurone grain; RW, residual wall (micrographs (a,e,f) Jacobsen et al.*, 1976; (b,c,d) Jacobsen and Pressman, unpublished data; courtesy of Springer-Verlag).*

consist of cells that are basically similar but molded into different shapes and located in various places relative to the embryo.

HORMONAL CONTROL OF ENDOSPERM CELL FUNCTION

When seed germination occurs, the endosperm is consumed by the enlarging embryo. In most cases the bulk of the endosperm is consumed after germination (radicle protrusion) has occurred. In wild oat (*Avenua fatua* L.), amylase production and therefore initiation of endosperm hydrolysis occurs at least 24 hr after germination has occurred (Chen and Varner, 1970; Chen and Chang, 1972). It appears that endosperm modification is largely a postgermination phenomenon and provides a nutritional support system to the enlarging embryo rather than initiating its enlargement. It seems likely that the same applies to other cereals. However, in seeds where the embryo is embedded in the endosperm, as in celery and lettuce, radicle protrusion cannot occur unless the endosperm is disrupted, and in such cases endosperm modification may be a requirement for germination. The important question in the present context is what initiates degradative changes in the endosperm.

Attempts to demonstrate embryonic control of endosperm behavior have not always yielded clear-cut results. In castor bean, gluconeogenesis in deembryonated seed occurs almost as well on imbibition of water as when the embryo is present (Huang and Beevers, 1974; Marriott and Northcote, 1975a,b) indicating only a small amount of embryonic control. However, application of GA_3 to deembryonated seed causes a considerable increase in the rate of lipolysis and in the rate of appearance and the final levels of some of the glyoxysomal and mitochondrial enzymes (Marriott and Northcote, 1975a,b; Wrigley and Lord, 1977). These studies provide some evidence of hormonal control of some of the gluconeogenetic processes in castor bean, but the control is far from complete and has not been shown to occur *in vivo*. The possibility exists that mature endosperm already contains gibberellin and thus exogenous gibberellin would exert only partial control. In contrast to castor bean, the appearance of glyoxysomal marker enzymes in wheat aleurone is completely dependent on the presence of the embryo, and the embryonic control can be replaced by GA_3 (Doig *et al.*, 1975). In barley aleurone, the studies of Jones and Price (1970) show that there are essentially no visible changes in cell structure in the absence of the embryo and that addition of GA_3 elicits all the responses already mentioned. As in wheat, gluconeogenesis in barley is probably also under hormonal control, although it has not been examined in any detail. Thus the major changes occurring in cereal aleurone during germination seem to be initiated hormonally.

Cell wall breakdown commonly occurs during endosperm modification. This process has occupied the attention of seed physiologists because the walls are usually thick and their breakdown is conspicuous, they represent a major carbohydrate storage pool in the endosperm, and they are primarily responsible for the physical integrity of the tissue. Thus their breakdown is of importance in embryo nutrition and in the penetration of the endosperm by the embryo. In fenugreek seeds, galactomannan breakdown, and production of enzymes (α-galactosidase, β-mannosidase and endo-β-mannanase) proceed in deembryonated endosperm as well as in the whole seed (Reid and Meier, 1972; Rijven, 1972; Reid and Meier, 1973; McCleary and Matheson, 1975; Reid and Davies, 1977) indicating that enzyme production is not under embryonic control, although endo-β-mannanase does not develop quite as well as in whole seed (Reid and Davies, 1977). Galactomannan breakdown in isolated endosperm was found to be enhanced by ethylene but not by GA_3, 2,4-D or kinetin. ABA was inhibitory, and this was counteracted by ethylene but not by GA_3, 2,4-D or kinetin (Rijven, 1972). Consequently, the evidence for hormonal control of cell wall breakdown is not convincing, although again the possibility of high background hormone levels has not been examined.

On the other hand, there are two cases where endosperm wall breakdown is under hormonal control. In isolated barley aleurone tissue, cell walls remain intact, but in the presence of GA_3, the walls break down extensively (Jones, 1969b; Jones, 1969c; Jones and Price, 1970; Taiz and Jones, 1970; Ashford and Jacobsen, 1974a). Similarly, the endosperm in deembryonated seed of celery does not seem to change, but in the presence of gibberellin A_{4+7}, the walls are modified in some way, presumably involving breakdown of some of the components, and the cells separate from each other (Jacobsen *et al.*, 1976).

In those cases where endosperm physiology can be shown to be hormonally controlled, the next question is whether hormonal control occurs *in vivo* and, if so, where the hormone comes from. Thus it is desirable to answer the following questions:

1. Which hormone can trigger the processes in isolated tissue?
2. Can the same hormone be found in the tissue *in vivo* and correlated with cellular changes?
3. Are the *in vivo* changes identical to those occurring in isolated endosperm treated with hormone?
4. Where does the hormone arise in the seed?

There is no case where all of these questions have been answered. The best evidence for hormonal control of endosperm function *in vivo* has come from cereal grains. An aleurone-activating factor from barley embryos was shown to be gibberellin-like (Yomo, 1960a,b) and to be synthesized in both the scutellum and in the nodal region of the embryonic axis (MacLeod and Palmer, 1966; Radley, 1967; 1969). The gibberellins, which appear to be GA_3 and GA_1 although they have not been unequivocally identified, are made following imbibition of water, and they move out of the embryo via the scutellum (Yomo and Iinuma, 1966). Gibberellins have been shown to replace embryonic control of endosperm hydrolase production (Paleg, 1960; Yomo, 1060c; Briggs, 1963; Paleg, 1964; Varner, 1964) and some endosperm structural changes (see earlier; Ashford and Jacobsen, 1974b). It has become widely accepted that the embryonic gibberellin is the sole cause of aleurone changes *in vivo*. While this may be true, it would seem wise not to regard it as being beyond doubt because there is evidence from wheat aleurone that other hormones can affect some processes. In wheat, lipid loss and lipase induction are regulated by IAA and kinetin but not by GA_3, whereas enzymes of gluconeogenesis seem to be under GA_3 control (Laidman *et al.*, 1974). Evidence for involvement of other hormones in similar aspects of barley aleurone function has not been possible to obtain (Newman and Briggs, 1976; Firn and Kende, 1974).

In summary, although there is considerable evidence for hormone responsiveness of mature endosperm tissue, the case for hormonal control of endosperm function during seed germination has been developed best in the case of barley grain. Although it seems likely that GA is a hormonal messenger between embryo and endosperm, the case is circumstantial, and whether it is the only hormone involved seems debatable. Equivalent evidence for hormonal embryo-endosperm communication in any other seed species has not been obtained.

ENZYME SYNTHESIS AND CONTROL

The demonstration of absolute control by GA_3 over some processes in barley aleurone has led to an extensive study of the nature of the response in the context of seed physiology (Yomo and Varner, 1971; Ho and Varner, 1976). Aleurone also has some attributes that make it particularly useful for study of the mechanism of hormone action. Barley aleurone consists basically of one cell type, and as far as can be told, the cells respond similarly and semisynchronously to GA_3. The cells neither grow, divide, nor differentiate during the response and aleurone can be obtained in large quantities with little effort.

The response of aleurone layers to GA_3 includes the release of a number of enzymes most of which are hydrolytic and which are responsible for starchy endosperm hydrolysis during germination of the grain. Because α-amylase is the most abundant of these enzymes and because it is stable, easily assayed, and easily purified, it has been studied in greatest detail. Amylase production was found to be inhibited by inhibitors of translation and transcription (see Yomo and Varner, 1971) and was later shown by direct methods to arise by new synthesis following GA_3 treatment (Varner and Chandra, 1964; Filner and Varner, 1967). In line with the new concepts of molecular biology of the time, the requirement for transcription led to the hypothesis that GA_3 might direct α-amylase synthesis by promoting synthesis of its mRNA (Varner, 1964). The rest of this section describes attempts to prove or disprove this hypothesis.

Initial studies were designed to determine what sort of RNA was necessary and whether its metabolism was affected by GA_3 in parallel with α-amylase synthesis. Studies with 5-fluorouracil, which preferentially inhibits synthesis of the major species of RNA, showed that even though synthesis of the major species was strongly suppressed, there was little effect on α-amylase synthesis (Jacobsen and Zwar, 1974). Furthermore, GA_3 had no effect on the total RNA level (Chandra and Varner, 1965; Jacobsen and Zwar, 1974) nor did it enhance accumulation of labeled precursors into total RNA or the major RNA species (Zwar and Jacobsen, 1972). Therefore it seemed likely that the important RNA was a minor species, and GA_3 was shown to enhance label accumulation into a polydisperse minor RNA fraction (Zwar and Jacobsen, 1972) that had some of the characteristics of mRNA, but template activity was not tested. Changes in this fraction began at about the same time as α-amylase synthesis. These results indicated that synthesis of the major RNA species was of little consequence to α-amylase induction but that synthesis of a minor fraction was not only necessary but correlated with α-amylase synthesis.

It seemed possible that this minor RNA fraction might be poly(A)-containing RNA. Consequently poly(A)-RNA was sought and identified in barley aleurone cells and its rate of synthesis and accumulation were shown to increase in the presence of GA_3 (Ho and Varner, 1974; Jacobsen and Zwar, 1974) commencing at about 3-4 hr after GA_3 application (Ho and Varner, 1974), which is about the same time as α-amylase synthesis. The synthesis of poly(A)-RNA was unaffected by 5-fluorouracil (Jacobsen and Zwar, unpublished data); therefore it was likely that poly(A)-RNA was the RNA species that was important for α-amylase synthesis.

To determine if poly(A)-RNA contained the mRNA for α-amylase, RNA from control and GA_3-treated aleurone layers was translated in a cell-free system prepared from wheat germ. A prominent GA_3-promoted polypeptide was identified as α-amylase on the basis of its similar, but not identical, molecular weight and its reaction with antiamylase antiserum (Higgins *et al.*, 1976). The profiles of *in vitro* synthesized products (Fig. 4) show that although the spectrum of polypeptides changed very little with time of incubation with GA_3, the proportions of the total radioactivity incorporated in α-amylase (upper arrow) and an unidentified polypeptide (lower arrow) increase with time. Since polypeptide synthesis is related to the amount of mRNA present (Higgins *et al.*, 1977) these results indicate that GA_3 caused an increase in the amount of translatable mRNA for α-amylase.

The increase in translatable mRNA with time after GA_3 treatment was quantified by expressing label incorporation into α-amylase as a percentage of the total label incorporation. The increase was related to the rate of amylase synthesis over the short- (Higgins *et al.*, 1976) and long term (Fig. 5a). Both α-amylase mRNA activity and α-amylase synthesis began to increase 2-3 hr after addition of GA_3, rose to a peak, and after

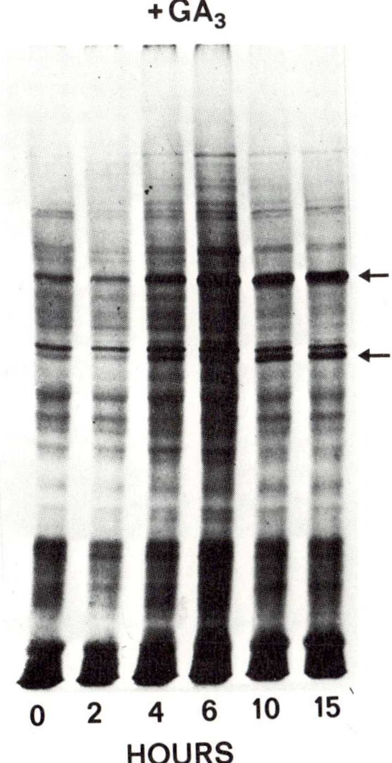

FIGURE 4. Fluorograph of an SDS-polyacrylamide gel showing profiles of polypeptides synthesized by the wheat germ system programmed with RNA isolated from barley aleurone layers incubated with GA_3 (10^{-6} M) for various periods of time. The top arrow indicates α-amylase and the bottom one an unidentified polypeptide.

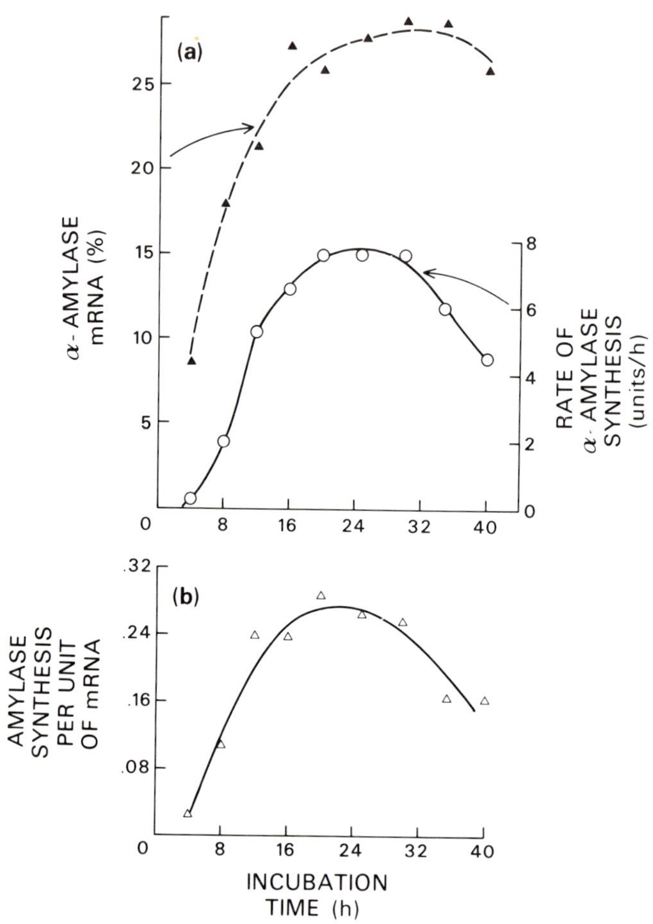

FIGURE 5. (a) Time course of α-amylase synthesis and the level of α-amylase mRNA in barley aleurone layers incubated with GA_3 (10^{-6}M). (b) Variation of the ratio of α-amylase synthesis to mRNA level (values taken from 5a) with time.

remaining constant for a while began to decline. But although there is a close relationship between amylase synthesis and translatable mRNA, their increases are not proportionate. When the curves of Fig. 5a are related by dividing the rate of amylase synthesis by the amount of translatable mRNA at each time point, it is found that as time progresses, amylase synthesis increases faster than translatable mRNA (Fig. 5b). One interpretation of these results is that the efficiency of translation of α-amylase mRNA increases relative to that of other messengers, but it is also possible that enzyme and mRNA activities do not reflect the true enzyme protein and mRNA levels. For example, in the early hours of enzyme synthesis, a proportion of the enzyme in the cells may be inactive, in which case we would underestimate efficiency. A similar underestimate would occur if some inactive form of α-amylase mRNA was activated during its isolation. It follows from these results that GA_3 may regulate α-amylase synthesis in two ways: (*a*) by increasing the level of translatable mRNA, and (*b*) by increasing the efficiency with which the mRNA is translated *in vivo*.

Abscisic acid antagonizes all of the GA_3-induced effects in barley aleurone tissue (Varner and Ho, 1976) including GA_3-induced amylase synthesis (Chrispeels and Varner, 1966) when added at the same time as GA_3. Recently, Ho and Varner (1976) found that ABA strongly inhibited amylase synthesis when added 12 hr after GA_3. At this time, amylase synthesis was no longer inhibited by inhibitors of transcription, indicating that RNA synthesis was no longer required (Ho and Varner, 1974) and the level of translatable mRNA for amylase was probably approaching its maximum (Higgins *et al.*, 1976). These results led to the deduction that ABA was not acting on transcription but on some posttranscriptional, perhaps translational, event. We have repeated these experiments in order to establish the relationship between α-amylase synthesis and its mRNA level.

Added at the same time as GA_3, ABA strongly suppressed the rate of amylase synthesis and the amount of translatable mRNA (Fig. 6) as well as the efficiency of mRNA translation (data

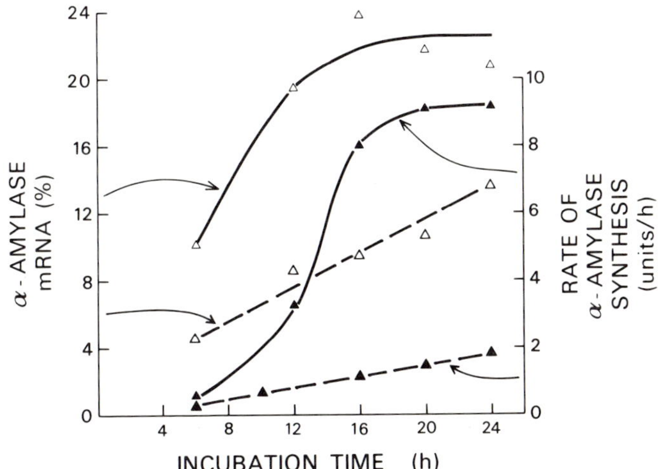

FIGURE 6. Suppression of α-amylase synthesis and the level of α-amylase mRNA in barley aleurone layers by ABA (2.5 x 10^{-5} M) added at the same time (0 h) as GA_3 (10^{-6} M). Solid lines are GA_3 only. Dotted lines are GA_3 + ABA.

not shown). Therefore, ABA once again suppressed the total GA_3 response. It does not seem to be an inhibitor of mRNA translation only. The results of midcourse addition of ABA are shown in Fig. 7. Added 7 and 12 hr after GA_3 when the level of translatable mRNA was still increasing (Fig. 7b), ABA prevented a further rise in both the rate of α-amylase synthesis (Fig. 7a) and the level of translatable mRNA (Fig. 7b).

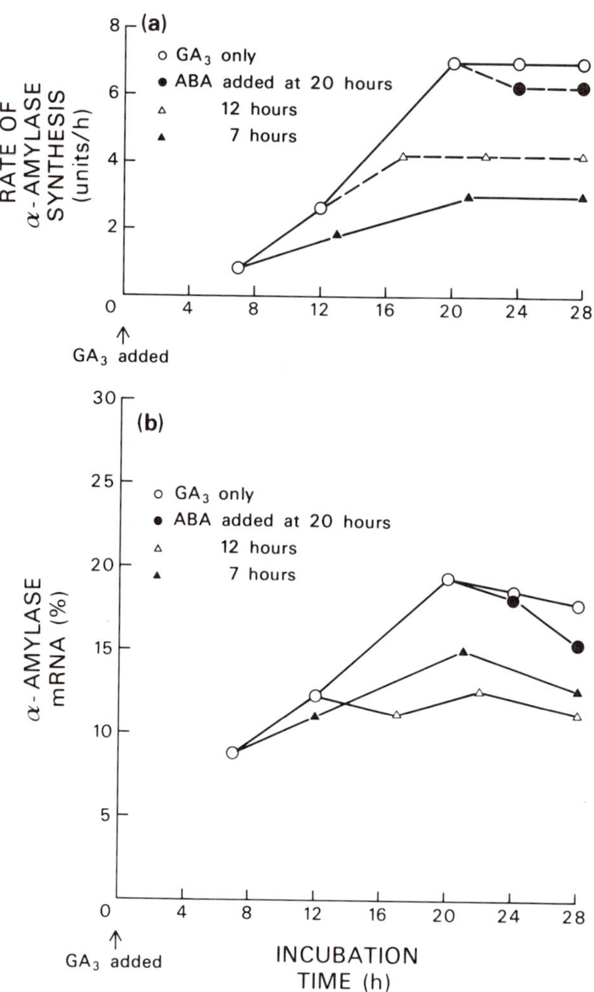

FIGURE 7. Effect of midcourse addition of ABA (2.5 x 10^{-5} M cistrans) on (a) α-amylase synthesis and (b) the level of α-amylase mRNA in barley aleurone layers. GA_3 (10^{-6} M) was present from time 0.

Notably, both attained a constant level, and over the course of the experiment neither dropped to zero. Added 20 hr after GA_3, when the level of translatable mRNA was at its maximum, ABA had only a small effect on both amylase synthesis and translatable mRNA level over the next 10 hr. The usual apparent rise in mRNA translational efficiency was also suppressed in line with mRNA activity and amylase synthesis (data not shown), except when ABA was added at 20 hr. Contrary to the results of Ho and Varner (1976), our results indicate that ABA cannot always rapidly suppress amylase synthesis. ABA has greatest effect before amylase mRNA level has maximized and even then it does not prevent amylase synthesis in the short term, but only prevents the rate of amylase synthesis from increasing further. The regulation of amylase synthesis by ABA seems to involve, and perhaps in part is mediated by, control of both the level of translatable α-amylase mRNA and its translatability.

As indicated above, studies of ABA action in aleurone to date have shown it to be a suppressor of GA_3-induced responses. However, our studies indicate that while ABA suppresses the increase in amylase mRNA, it enhances the levels of some other mRNA species. Fig. 8 shows that when ABA was added 7 hr after GA_3, the levels of five polypeptides synthesized *in vitro* were enhanced within an additional 6 hr (see frame 7/13), presumably reflecting similar increases in their mRNAs. Similar results were obtained when ABA was added 12 or 20 hr after GA_3, at the same time as GA_3, and in the absence of GA_3, so that stimulation of *in vitro* protein synthesis appears to be an independent effect of ABA. Furthermore, new proteins were synthesized *in vivo* in the presence of ABA (Fig. 9). These effects of ABA could reflect synthesis of structural or enzymatic proteins or alternatively they may have some significance in the regulation of α-amylase synthesis. In any case, ABA appears to be both a suppressor and an inducer of mRNA and protein levels in aleurone cells.

The study of mRNA in relation to GA_3 and ABA action has permitted new insight into hormonal control of protein synthesis in aleurone cells, although the results should be interpreted cautiously. In the studies described above, only the activities of mRNAs for α-amylase and other proteins have been quantified, and such activities do not necessarily reflect the total amount of mRNA, some of which may exist in inactive form(s). Therefore, any conclusions drawn from these studies are tentative and require confirmation by studies in which the total amount of mRNA is determined.

The intensive study of hormone action in aleurone has been motivated primarily by the thought that in this tissue it may be possible to determine the mechanism of GA_3 and ABA action. However, the structural and physiological similarities between

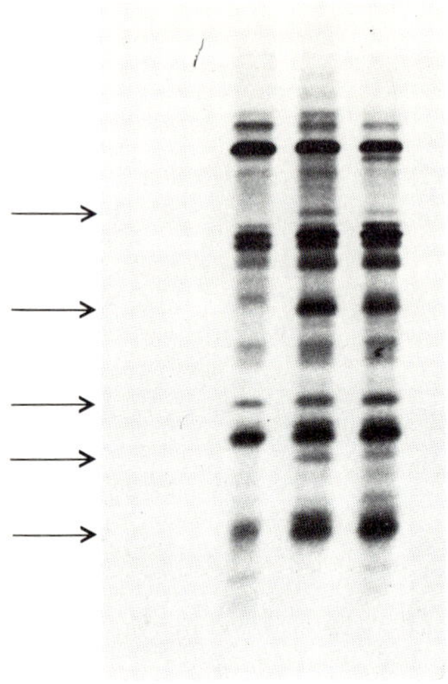

C 7 7/13 7/21

FIGURE 8. Fluorograph of an SDS-polyacrylamide gel showing the electrophoretic separation of polypeptides synthesized by the wheat germ system programmed with mRNA isolated from barley aleurone layers incubated with GA_3 (10^{-6} M) and ABA (2.5×10^{-5} M cis-trans). C, no mRNA added; 7, incubated with GA_3 for 7 hr; 7/13, incubated with GA_3 for 13 hr and ABA present from 7 to 13 hr; 7/21, incubated with GA_3 for 21 hr and ABA present 7 to 21 hr. The arrows point to ABA-induced polypeptides.

aleurone cells and the cells of other living endosperm tissues present the possibility that in the study of aleurone, one is studying the physiology of a common living endosperm cell type. Thus it is of interest to compare the responses of other endosperm cells to GA_3 and ABA with those of aleurone.

SUMMARY

Endosperm is the major storage tissue in the mature seeds of many plant species. It commonly consists of living cells, but in some cases, as in the cereal grains, it is composed largely of dead cells. Although only a few cases have been examined, living endosperm cells in different plant families have a number of structural features in common. Furthermore,

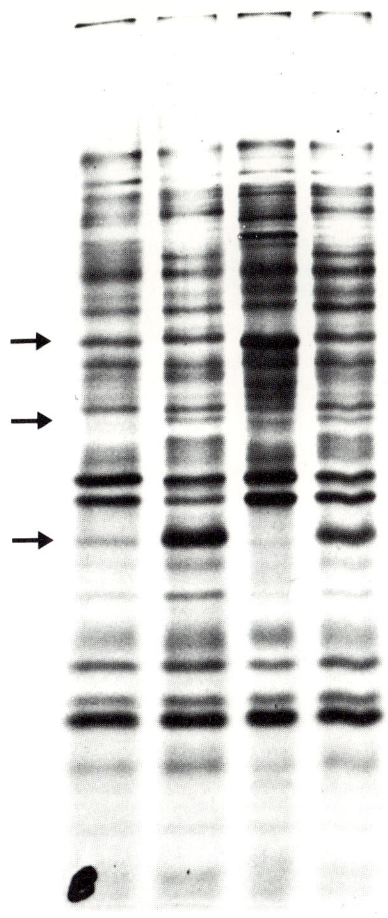

FIGURE 9. Fluorograph of an SDS-polyacrylamide gel showing profiles of polypeptides synthesized by barley aleurone layers in the absence (C) and presence of GA_3 (10^{-6} M) and ABA (2.5×10^{-5} M cis-trans). Hormones were present for 7 hr and label (^{35}S-methionine) was added for the last hour. The upper arrow points to α-amylase and the lower two arrows to two prominent ABA-induced polypeptides (see ABA and ABA + GA_3 columns).

during germination and endosperm breakdown similar structural changes occur in these cells.

It has been possible to establish embryonic control of endosperm breakdown in some seeds but only partial control or none at all in others. Where control is complete, the embryo can be replaced by gibberellin at least in part. Hormones have only small effects where incomplete or no control occurs. Thus endosperm function can be hormonally controlled in some cases, but whether this occurs *in vivo* and whether gibberellin is the only hormone involved is not yet clear.

Endosperm response to gibberellin has been examined in greatest depth in the case of barley aleurone. Among other events is the synthesis of new enzymes, including α-amylase, for which RNA synthesis (probably mRNA) is required. The mRNA for α-amylase from aleurone has been quantified by *in vitro* translation and its level shown to be modulated by gibberellin (GA_3) and abscisic acid (ABA) in a manner similar to α-amylase synthesis. However, the efficiency of mRNA translation also appears to increase with GA_3 treatment. Therefore, the level of mRNA and its translation efficiency seem to be major elements in the control of α-amylase synthesis.

In these studies, ABA acted not only as a suppressor of α-amylase synthesis and α-amylase mRNA level, but also as an inducer of apparently new protein and mRNA species.

REFERENCES

Ashford, A. E., and Jacobsen, J. V. (1974a). Cytochemical localization of phosphatase in barley aleurone cells: The pathway of gibberellic-acid-induced enzyme release. *Planta (Berl.) 120*, 81-105.

Ashford, A. E., and Jacobsen, J. V. (1974b). Cytochemical localization of acid phosphatase in isolated aleurone layers treated with gibberellic acid and in germinating barley grains. *In* "Mechanisms of Regulation of Plant Growth" (R. L. Bieleski, A. R. Ferguson, and M. M. Cresswell, eds.), Bulletin 12, pp. 591-599. The Royal Society of New Zealand, Wellington.

Ashton, F. (1976). Mobilization of storage proteins in seeds. *Ann. Rev. Plant Physiol. 27*, 95-117.

Briggs, D. E. (1963). Biochemistry of barley germination. Action of gibberellic acid on barley endosperm. *J. Inst. Brew., London 69*, 13-19.

Chandra, G. R., and Varner, J. E. (1965). Gibberellic acid controlled metabolism of RNA in aleurone cells of barley. *Biochim. Biophys. Acta 108*, 583-592.

Chen, S. S. C., and Chang, J. L. L. (1972). Does gibberellic acid stimulate seed germination via amylase synthesis? *Plant Physiol. 49,* 441-442.

Chen, S. S. C., and Varner, J. E. (1970). Respiration and protein synthesis in dormant and after-ripened seeds of *Avena fatua. Plant Physiol. 46,* 108-112.

Ching, T. M. (1972). Metabolism of germinating seeds. *In* "Seed Biology," Vol. 11 (T. T. Kozlowski, ed.), Academic Press, New York.

Chrispeels, M. J., and Varner, J. E. (1966). Inhibition of gibberellic acid induced formation of α-amylase by abscisin 11. *Nature 212,* 1066-1067.

Corner, E. J. H. (1951). The leguminous seed. *Phytomorphol. 1,* 117-150.

Doig, R. I., Colborne, A. J., Morris, G., and Laidman, D. L. (1975). The induction of glyoxysomal enzyme activities in the aleurone cells of germinating wheat. *J. Exp. Bot. 26,* 387-398.

Eb, van der, A. A., and Niewdorp, P. J. (1967). Electron microscope structure of the aleurone cells of barley during germination. *Acta bot. Neerl. 15,* 690-699.

Filner, P., and Varner, J. E. (1967). A simple and unequivocal test for *de novo* synthesis of enzymes: Density labeling of barley α-amylase with H_2O^{18}. *Proc. Nat. Acad. Sci. (U.S.) 58,* 1520-1526.

Firn, R. D., and Kende, H. (1974). Some effects of applied gibberellic acid on the synthesis and degradation of lipids in isolated barley aleurone layers. *Plant Physiol. 54,* 911-915.

Fulcher, R. G. (1972). Observations on the aleurone layer with emphasis on wheat. Ph.D. dissertation, Monash Univ., Clayton, Victoria, Australia.

Higgins, T. J. V., Zwar, J. A., and Jacobsen, J. V. (1976). Gibberellic acid enhances the level of translatable mRNA for α-amylase in barley aleurone layers. *Nature 260,* 166-169.

Higgins, T. J. V., Zwar, J. A., and Jacobsen, J. V. (1977). Hormonal control of the level of translatable mRNA for α-amylase in barley aleurone cells. *In* "Acides nucleiques et synthese des proteines chez les vegetaus." Colloques Internationaux C.N.R.S. (J. H. Weil and L. Bogorad, eds.), No. 261, pp. 481-486. Paris.

Ho, D. T. H., and Varner, J. E. (1974). Hormonal control of messenger ribonucleic acid metabolism in barley aleurone layers. *Proc. Nat. Acad. Sci. (U.S.) 71,* 4783-4786.

Ho, D. T. H., and Varner, J. E. (1976). Responses of barley aleurone layers to abscisic acid. *Plant Physiol. 57,* 175-178.

Huang, A. H. C., and Beevers, H. (1974). Developmental changes in endosperm of germinating castor bean independent of embryonic axis. *Plant Physiol. 54*, 277-279.

Jacobsen, J. V., and Zwar, J. A. (1974). Gibberellic acid and RNA synthesis in barley aleurone layers: Metabolism of rRNA and tRNA and RNA containing polyadenylic acid sequences. *Aust. J. Plant Physiol. 1*, 343-356.

Jacobsen, J. V., Knox, R. B., and Pyliotis, N. A. (1971). The structure and composition of aleurone grains in the barley aleurone layer. *Planta (Berl.) 101*, 189-209.

Jacobsen, J. V., Pressman, E., and Pyliotis, N. A. (1976). Gibberellin-induced separation of cells in isolated endosperm of celery seed. *Planta (Berl.) 129*, 113-122.

Jones, R. L. (1969a). The fine structure of barley aleurone cells. *Planta (Berl.) 85*, 359-375.

Jones, R. L. (1969b). Gibberellic acid and the fine structure of barley aleurone cells. I. Changes during the lag-phase of α-amylase synthesis. *Planta (Berl.) 87*, 119-133.

Jones, R. L. (1969c). Gibberellic acid and the fine structure of barley aleurone cells. II. Changes during the synthesis and secretion of α-amylase. *Planta (Berl.) 88*, 73-86.

Jones, R. L. (1974). The structure of the lettuce endosperm. *Planta (Berl.) 121*, 133-146.

Jones, R. L., and Price, J. M. (1970). Gibberellic acid and the fine structure of barley aleurone cells. III. Vacuolation of the aleurone cell during the phase of ribonuclease release. *Planta (Berl.) 94*, 191-202.

Laidman, D. L., Colborne, A. J., Doig, R. I., and Varty, K. (1974). *In* "Mechanisms of Regulation of Plant Growth" R. L. Bieleski, A. R. Ferguson, and M. M. Creswell, eds.), Bulletin 12, pp. 581-590. The Royal Society of New Zealand, Wellington.

MacLeod, A. M., and Palmer, G. H. (1966). The embryo of barley in relation to modification of the endosperm. *J. Inst. Brew., London,* 580-589.

McCleary, B. V., and Matheson, N. K. (1975). Galactomannan structure and β-mannanase and β-mannosidase activity in germinating legume seeds. *Phytochem. 14*, 1187-1194.

McNeil, M., Albersheim, P., Taiz, L., and Jones, R. L. (1975). The structure of plant cell walls. VII. Barley aleurone cells. *Plant Physiol. 55*, 64-68.

Marriott, K. M., and Northcote, D. H. (1975a). The breakdown of lipid reserves in the endosperm of germinating castor beans. *Biochem. J. 148*, 139-144.

Marriott, K. M., and Northcote, D. H. (1975b). The induction of enzyme activity in the endosperm of germinating castor bean seeds. *Biochem. J. 152*, 65-70.

Martin, A. C. (1946). The comparative internal morphology of seeds. *Amer. Midland Naturalist 36*, 573-660.

Millerd, A. (1975). Biochemistry of legume seed proteins. *Ann. Rev. Plant Physiol. 26,* 53-72.

Mollenhauer, H. H., and Totten, C. (1970). Studies on seeds. V. Microbodies, glyoxysomes, and ricinosomes of castor bean endosperm. *Plant Physiol. 46,* 794-799.

Netolitzky, F. (1926). Anatomie der Angiospermen-Samen. *Handbuch der Pflanzenanatomie.* Bd. X. Berlin: Borntraeger.

Newman, J. C., and Briggs, D. E. (1976). Glyceride metabolism and gluconeogenesis in barley endosperm. *Phytochem. 15,* 1453-1458.

Paleg, L. (1960). Physiological effects of gibberellic acid. 1. On carbohydrate metabolism and amylase activity of barley endosperm. *Plant Physiol. 35,* 293-299.

Paleg, L. (1964). Cellular localization of the gibberellin-induced response of barley endosperm. *In* "Regulateurs naturels de la croissance vegetale" (J. P. Nitsch, ed.), pp. 303-317. C.N.R.S. Paris.

Paleg, L., and Hyde, B. (1964). Physiological effects of gibberellin acid. VII. Electron microscopy of barley aleurone cells. *Plant Physiol. 39,* 673-680.

Radley, M. (1967). Site of production of gibberellin-like substances in germinating barley embryos. *Planta 75,* 164-171.

Radley, M. (1969). The effect of the endosperm on the formation of gibberellin by barley embryos. *Planta 86,* 218-223.

Reid, J. S. G., and Davies, C. (1977). Endo-β-mannanase, the leguminous aleurone layer and the storage galactomannan in germinating seeds of *Trigonella foenum-graecum* L. *Planta (Berl.) 133,* 219-222.

Reid, J. S. G., and Meier, H. (1972). The function of the aleurone layer during galactomannan mobilization in germinating seeds of fenugreek *(Trigonella foenum-graecum* L.*)*, crimson clover *(Trifolium incarnatum* L.) and lucerne *Medicago sativa* L.): A correlative biochemical and ultrastructural study. *Planta (Berl.) 106.* 44-60.

Reid, J. S. G., and Meier, H. (1973). Enzymic activities and galactomannan mobilisation in germinating seeds of fenugreek *(Trigonella foenum-graecum* L. *Leguminosae).* Secretion of α-galactosidase and β-mannosidase by the aleurone layer. *Planta (Berl.) 112,* 301-308.

Rijven, A. H. G. C. (1972). Control of the activity of the aleurone layer of fenugreek, *Trigonella foenum-graecum* L. *Acta Bot. Neerl. 21,* 381-386.

Seiler, A. (1977). Galaktomannanabbau in keimenden Johannisbrotsamen *(Ceratonia siliqua* L.*) Planta (Berl.) 134,* 209-221.

Taiz, L., and Honigman, W. A. (1976). Production of cell wall hydrolyzing enzymes by barley aleurone alyers in response to gibberellic acid. *Plant Physiol. 58,* 380-386.

Taiz, L., and Jones, R. L. (1970). Gibberellic acid, β-1, 3-glucanase and the cell walls of barley aleurone cells. *Planta (Berl.) 92,* 73-84.

Tully, R. E., and Beevers, H. (1976). Protein bodies of castor bean endosperm. Isolation, fractionation, and the characterization of protein components. *Plant Physiol. 58,* 710-716.

Varner, J. E. (1964). Gibberellic acid controlled synthesis of α-amylase in barley endosperm. *Plant Physiol. 39,* 413-415.

Varner, J. E., and Chandra, G. R. (1964). Hormonal control of enzyme synthesis in barley endosperm. *Proc. Nat. Acad. Sci. (U.S.) 52,* 100-106.

Varner, J. E., and Ho, D. T. H. (1976). The role of hormones in the integration of seedling growth. *In* "The Molecular Biology of Hormone Action" (J. Papaconstantinou, ed.), pp. 173-194. Academic Press, New York.

Vigil, E. L. (1970). Cytochemical and developmental changes in microbodies (glyoxysomes) and related organelles of castor bean endosperm. *J. Cell Biol. 46,* 435-454.

Wrigley, A., and Lord, J. M. (1977). The effects of gibberellic acid on organelle biogenesis in the endosperm of germinating castor bean seeds. *J. Exptl. Bot. 28,* 345-353.

Yomo, H. (1960a). Studies on the amylase activating substance (part 2) on the amylase activating substance in the embryo culture medium and the barley malt extract. *Hakko Kyokai Shi 18,* 494-499.

Yomo, H. (1960b). Studies on the amylase activating substance (part 5) purification of the amylase activating substance in barley malt and its properties. *Hakko Kyokai Shi 18,* 603-606.

Yomo, H. (1960c). Studies on the amylase activating substance (part 4) on the amylase activating action of gibberellin. *Hakko Kyokai Shi 18,* 600-603.

Yomo, H., and Iinuma, H. (1966). Production of gibberellin-like substance in the embryo of barley during germination. *Planta 71,* 113-118.

Yomo, H., and Varner, J. E. (1971). Hormonal control of a secretory tissue. *In* "Current Topics in Developmental Biology" (A. Moscona and A. Monroy, eds.), Vol. 6, pp. 111-114. Academic Press, New York.

Youle, R. J., and Huang, A. H. C. (1976). Protein bodies from the endosperm of castor bean. Subfractionation, protein components, lectins, and changes during germination. *Plant Physiol. 58,* 703-709.

Zwar, J. A., and Jacobsen, J. V. (1972). A correlation between a ribonucleic acid fraction selectively labelled in the presence of gibberellic acid and amylase synthesis in barley aleurone layers. *Plant Physiol. 49,* 1000-1006.

Index

F

G

H

I

K

L

M

T

V

W

X

Z